THE NAKED DARWIN

A SERIOUS LOOK AT AN AGING IDEA

JOHN SIMPSON

AR PRESS

For more about this author please visit www.thenakeddarwin.com

Paperback ISBN: 978-1-965092-41-5
Hardcover ISBN: 978-1-965092-42-2

1. Books › Science & Math › Evolution › General
2. Books › Science & Math › History & Philosophy › History of Science

Second Edition Revised

AR PRESS

Published by: AR PRESS
Roger L. Brooks, Publisher
roger@americanrealpublishing.com
americanrealpublishing.com

For Shelley

PREFACE TO THE SECOND EDITION

In this revised edition of *The Naked Darwin*, I have made slight changes to the text of the first edition for the purpose of clarification. I have eliminated an unnecessary chapter and added the Addendum for those readers who would like more detail on the problems with Darwin's ideas today.

ACKNOWLEDGMENTS

I owe my heartfelt thanks to many friends who helped make this book possible. Thanks to Myra Epping for reading my rough draft and devoting so much labor to initial improvements. Thanks to John Petter for additional help and instructions. Thanks to Duane Young for his critical thoughts, editing, and for the many books over the years. Thanks to my brother Marc for his help and for putting wind in my sails. A special thank you goes to my band of brothers: Greg Butler, Dennis Eldridge, Robert Lackie, and Tim Rice. I could not have done it without your help. Thanks to Polly Danforth for your labor and encouragement; you went above and beyond for this. And thanks to Phillip E. Johnson for planting the first seeds of doubt.

...his notebooks make plain that competition, free trade, imperialism, racial extermination, and sexual inequality were written into the equation from the start—'Darwinism' was always intended to explain human society.

From Darwin,
by Adrian Desmond and James Moore, 1991, p. xxi.

PREFACE

IT IS SATURDAY MORNING AND I have chores to do: got to clean up the basement. Want to leaven the work with music. Take my radio and tune in to the golden oldie station, but golden oldies are not cutting it this morning. This cleanup is going to take more than an hour. I have time for what I really love. I ditch the golden oldies, head upstairs to borrow my son's CD player and grab a CD of Mahler's Third Symphony. Set up, plug in, the unison horns boom out the first notes. Damn, what a way to start! I am already transported from the basement. I grab the broom and the dustpan, and it does not matter. Does not matter that I am in my grubby basement sweeping up years of dirt and debris because I am only physically in my basement. My soul, my spirit, name your own term, has been transported by Mahler's world of sound. I hardly notice the work. I am on a musical journey that is one of the deepest experiences any one of us can have.

I am betting you have had something like that happen to you. Maybe for you it is Aretha Franklin, or James Taylor, or Big Band Jazz, or Puccini, or Louis Armstrong or even the late quartets of Beethoven. You have experienced the power of music. And not just music, maybe you are moved by the beauty of great prose or poetry or the well-constructed essay. There are all manner of beautiful things in this world that affect us deeply. How can we explain it? Can this sort of experience be *explained*? Do we even care whether or not it *is* explained? These beauties are part of the best life has to offer us, and they raise a question in our minds about the explanation of our world bequeathed to us by Charles Darwin. The profound impact that art has on us creates in us a nagging doubt that natural selection working by unguided processes could produce Louis Armstrong's "Potato Head Blues," or Puccini's "Nesum Dorma," or Eric Clapton's solos. Not only do we wonder how such things could exist by such a process, but we also wonder how could it be that we are so deeply touched by them. Now I know that Darwin and his modern-day defenders tell us how, but even those who accept the theory's explanation of how the giraffe came to be find the theory inadequate for explaining Paul Simon's "Bridge Over Troubled Waters." And then there are many of you who have your doubts about the giraffe as well. This book begins with doubts about the theory of evolution and its purpose is to encourage doubt and justify skepticism.

I write as one nonscientist to others about the state of the debate on Darwin's theory here at the beginning of the twenty-first century. On the subject of evolution, I am self-educated and no expert. I hope that my experience and thoughts will be useful to those wanting to better understand the controversy and trying to make sense of the noise from the popular press.

Darwin published his famous book in 1859 and it can be truly said that his ideas rank with the most influential in the history of Western thought. Many do not think of his legacy as theory at all, but plain and simple fact. The first time I read a serious critique of Darwin I hardly knew what to make of it. The writer certainly knew the subject and made his points well, but geesh! Darwinian evolution, not accurate? I wondered, *Could it be possible for an intelligent educated person to question his theory?*

The critique I read simply bounced off my mind. There was no place in my thinking for the critique to gain a purchase. I put the book down and did not give it serious thought for a few years. But doubts began to grow. How could the theory of evolution explain what Mahler's music did to me? And if it was inadequate to explain Mahler—the theory is said to account for everything that there is—maybe the critic was right. Maybe it was inadequate to explain biological life as well.

In large part my reaction to the critique involved thoughts such as, *this cannot be so because it would mean that science, for the last one hundred plus years, has in some major way got a large part of the picture wrong*. Science wrong? Scientists, wrong?! …nah. Not possible. Not that many of them. Not for this many years. How could it be?

Another reason I could not accept the critique was due to an idealized picture of how science gets done. In my naiveté I believed scientists were the most objective of humans, caring only for the truth. They worked in labs unmoved by social forces or politics. They worked according to something called the "scientific method," which insured that their pronouncements were true. Scientists were not subject to the failings and weaknesses of the rest of us, at least not while doing science. Jealousy, the need for approval, vindictiveness, insecurities, peer pressure—in short, things that mar my life and yours to one degree or another—did not enter into the enterprise I thought of as science. The great brains of science, which thought and cared only for truth, would not, and could not, allow any petty weaknesses to cloud their conclusions.

I have been disabused of such simplistic notions. I was naïve. I should have known better. Scientists are human. Part of my own discovery was in understanding that scientists are not immune to the human failings that affect us all. Peer pressure and jealousy do enter into the scientific enterprise and sometimes do affect results. Scientists do not operate in an insulated community where truth can be pursued with purity. As we will learn later, Darwin's own ideas were born in a certain social and political climate and took hold, in part, for reasons that had nothing to do with their merit.

But take hold they did, and to such a degree and in such a manner that for many it is nearly inconceivable that Darwin's ideas can be challenged. I understand and sympathize. I was not too far from being in the same place myself. But then I could not shake the doubt that had taken root in my thinking. Is it possible? Could it be? The possibility for even a little doubt was crucial. Might the theory of evolution be a bit oversold? For many, there is simply no place for such doubt or questioning; for them it is not possible to doubt this theory in any way. The idea of Darwinian evolution for the devoted is not just settled, like the idea that the earth revolves around the sun, but it is for them something akin to religion. The theory means so much, is so important to their whole view of life, it is no longer science but ideology from which there is no retreat.

For those whose vocation is science related, whose careers are connected to Darwinian evolution, one can only imagine how difficult it must be to entertain doubts. Think of the person teaching science in a university setting where Darwinian evolution is the foundation of all that they teach. Then think what it would mean for them in midlife to consider that the basis for their view of the world is faulty. That would be huge. So much will be lost if they allow the slightest doubt. Not only would they jeopardize their career and their standing with their peers, but most devastatingly they would have to admit that they were wrong about the very basis of their career: a very, very difficult thing to do. We cannot expect very many to do so. Keeping all of this in view will help to understand the tenor of the debate today.

If you have the capacity to doubt some of what you have been told about how living things came to be, even just a little bit, the following may be helpful. Even if your doubt is not a firmly formed thought such as, "How could the first cell come into being by nothing more than chemical processes?" or "how did nonlife bring about life?" or "how could a reptile evolve into a bird?" Even if you just have nothing more than a vague sense that the Darwinian view of the world may not be able to account for all of reality as you know it, you may want to read on. In the following pages we will look at the story of how Darwin's *Origin* came to be, at how his theory came to dominate the natural sciences, at the evidence for and against (if it is a theory, there can be evidence both pro and con) and take a brief look at a possible challenger to Darwinian thinking. Most of all, I want to feed your doubts and bolster your skepticism.

TABLE OF CONTENTS

CHAPTER 1

BEFORE THE JOURNEY

Nurs'd by warm sun-beams in primeval caves
Organic life began beneath the waves…
Hence without parent by spontaneous birth
Rise the first specks of animated earth.

—Erasmus Darwin, *Desmond and Moore,* 1991, p. 5.

THE MERITS OF DARWIN'S REVOLUTIONARY ideas must stand or fall on the strength of the evidence. It is the evidence that will matter most for thinking people. At the same time, it is only fair to point out that, however strong a case Darwin made for his ideas, in 1859, his case was largely circumstantial. How could it be otherwise? Darwin claimed to tell us what happened thousands, no millions of years ago. Although he could record changes within a species and experiment with his pigeons, he could not show a single species evolving into another. In the laboratory he could not demonstrate the step-by-step creation of the eye. A large part of what went into it was observation followed by extrapolation. Moreover, great men of science in his own day such as Richard Owen, Louis Agassiz, François Jules Picet, William Hopkins, and Fleeming Jenkin saw the same natural phenomena that he did and drew different conclusions.

Darwin's theory, unlike Einstein's theory of special relativity, could not be tested by experiment and mathematics. With only a theoretical notion of how species came to be and how life originated, we find ourselves on very different ground. A consideration of the influences on Darwin's life helps us evaluate the strength of his ideas as well as understand possible origins of them. We do not seek to denigrate the ideas he advanced, but to state a simple but necessary truth, that in fairness should be a part of the discussion

The idea that nature, unaided by an outside intelligence, had produced all the panoply of life was not new with Charles Darwin; it had had its adherents since the time of ancient Greek civilization.[1] Over the years it has had its proponents and opponents. At the turn of the nineteenth century, this idea was a hot topic, but one championed by the radicals in Darwin's society. It was at that time articulated in the "penny press." Many of those who held such a view did so for very nonbiological reasons, reasons that had nothing to do with natural selection or any theory that would explain the diversity of life. The radicals of Darwin's day wanted to wrest political power from the establishment. That meant breaking a hold on society maintained by the Church of England and the network formed by those who graduated from Oxford and Cambridge.

Darwin would take what was radical and make it respectable. He would do this by providing a scientific underpinning for an idea that had been on the fringe in his own day. Though he was not a Christian, he was part of the landed gentry and had studied at Cambridge. In today's language, Darwin was a part of the establishment. By his theory of natural selection, he would make the scandalous respectable, and would claim to show that nature was the whole show, with no need of God to account for the origin of species. It was a stunning achievement. Darwin turned the direction of Western civilization like few have ever done.

On the same day, February 12, 1809, that Abraham Lincoln was born into poverty in rural Kentucky, Charles Darwin was born into wealth in rural Shropshire. His family's wealth made it possible for him to devote his life to his love of natural science without the necessity of earning a living. Charles was also born into a family of freethinkers. English society was dominated by the Anglican Church at this time, but his immediate and extended family would provide him a nontraditional, nonorthodox home environment by the standards of his day. Before Charles was born, his grandfather Erasmus Darwin had already ceased

1 Michael Denton, *Evolution: A Theory in Crisis,* Adler & Adler, Chevy Chase, MD, 1985, p. 37.

worshipping in church and was worshipping nature. He had even written his own book on evolution, *Zoonomia*, that set forth his ideas of how the natural world came to be through nature alone with no need for any creating God.

The Darwin's were connected by marriage to the Wedgwood family. Charles's father-in-law, Josiah Wedgwood, of pottery fame, was an early leader in England's Industrial Revolution. Though not as radical in his thinking as Erasmus, Josiah Wedgwood was a Unitarian, and as such had given up on orthodox views of Christ and the Trinity. Although this may not seem radical to a reader today, to grow up under such influences in England in the first half of the nineteenth century was to live in a world that challenged the status quo. What Charles received from his family intellectually and theologically helped make it possible for him to entertain unorthodox and even radical ideas with much greater ease than it would have been had he grown up in a traditional Church of England environment. Because his father wanted his son to be a medical doctor like himself, the teenage Charles Darwin was sent to one of the most radical city settings in the UK at that time. His experience as a teenager in Edinburgh would further move him along a path of rejecting traditional concepts of the natural world.

Darwin's grandfather and father were both Doctors of Medicine, and both had studied medicine in Edinburgh. So, at the age of sixteen in the summer of 1825, Charles was sent to the same city to follow in their footsteps. Edinburgh was a vibrant cosmopolitan city that Desmond and Moore describe as "...a city of wonderful sights and fearful sciences: of socialists experimenting with cooperative living, and...of professors debating the earth's origin, and anatomists debunking the creation of life."[2]

The following year Charles was inducted into one of several of Edinburgh's radical student societies, the Plinian. In 1826, the year Darwin joined, "...it had been penetrated by radical students—fiery, freethinking democrats who demanded that science be based on physical causes, not supernatural forces."[3] Darwin was proposed for the Plinian by William Browne. "Radicals like him were intent on reforming the

---

2 Adrian Desmond and James Moore, *Darwin,* W. W. Norton & Co., New York London, 1991, p. 23.

3 Desmond and Moore, *Darwin,* p. 31.

Church-dominated society. The Established Churches of Scotland and England ruled all aspects of life, monopolizing political offices, regulating hospital, university and legal posts …"[4] The week following his election Darwin heard a paper read that proposed that "The lower animals possess every faculty and propensity of the human mind."[5] The speaker, a Unitarian, "…had learned to see nature purely in terms of physical forces …"[6] Darwin's Edinburgh mentor would be the radical Robert Edmond Grant. According to Desmond and Moore, "Theirs was a decisive meeting. Darwin was coming under the wing of an uncompromising evolutionist …" who believed that "The origin and evolution of life were due simply to physical and chemical forces, all obeying natural laws."[7]

Here was an impressionable young man, away from home for the first time, caught up in the latest intellectual thinking of his day—nature with no need of God. He was invited to join with those who wanted to challenge the established order by knocking God out of it, and assaulting man's unique place in the natural order. And here was a young man well suited to give such thinking a welcome hearing. Other men at his age hearing such talk might have been shocked, but Darwin's upbringing prepared him to take such talk in stride. At this impressionable point in his life, he was caught up with those who believed that an evolutionary view of nature was true and God was not necessary. Here in Edinburgh, as a teenager, the trajectory of his life was being set. Long before he sailed on HMS Beagle, he was beginning to reject the orthodox view of origins and seriously consider the radical view that all creation was the result of natural causes.

Add to this Darwin's way of seeing nature. His view of the natural world came to him through his fascination with the lower orders of animals. At Edinburgh, his mentor Edmond Grant would share with Darwin his love of sponges; and during his Cambridge years, Charles was known for his own love of beetles. Many naturalists of his day looked at the higher orders of animals and saw beauty and grandeur. They could point

4 Desmond and Moore, *Darwin*, p. 32.

5 Desmond and Moore, *Darwin*, p. 32.

6 Desmond and Moore, *Darwin*, p. 32.

7 Desmond and Moore, *Darwin*, p. 34.

to the terrible beauty of the eagle. The animal displays strength, grace, and power. For them, the eagle could reveal the design of a great and wise creator. Darwin's eye was not looking up to the eagle, but down to beetles and sponges. His focus was on a different sort of creature. He had his nose in the weeds along the banks of streams and in the shallow pools at the ocean's edge. This focus helped produce in him a different view of what nature was all about. When he contemplated nature, he did not see beauty and purpose, but waste and cruelty. To his friend Joseph Hooker he wrote in 1856, "What a book a Devil's Chaplain might write in the clumsy, wasteful, blundering low & horridly cruel works of nature."

We know that the way a person looks at the world has a lot to do with what he or she sees in the world. All of us view life through our own set of lenses. Both the eagle and the beetle are a part of nature. A person looking up at the eagle soaring above may draw one kind of lesson from nature; the person looking down at beetles struggling to devour one another may draw an opposite lesson. From his early days, Darwin was one who was looking down and saw there a cruel struggle for survival. In watching beetles, he must have attributed to them a sentience something like that of higher animals and believed that they suffered to some degree as people might suffer. This way of seeing and feeling the pain of lower creatures seems a crucial aspect of what moved him to think and come to believe as he did. And this outlook intensified as he grew older. For him, nature was a cruel war and a brutal struggle for survival, not a world of beauty and harmony.

CHAPTER 2

AROUND THE WORLD

...owing to Darwin, our entire picture of man and of his place and role in nature has been transformed... As regards man it became ever clearer that he had originated from an ancestor sufficiently ape-like to be placed in the same group with the existing great apes.

—From the Introduction to the Mentor edition of *The Origin of Species,* Sir Julian Huxley.

TWO DAYS AFTER CHRISTMAS IN 1831, Charles Darwin sailed on HMS Beagle for South America. Many of us know that this is one of the most momentous voyages for the history of Western thought. Darwin sailed at age twenty-two, and returned five years later having experienced sights and sounds from another world. He returned with a head full of questions but was convinced of the truth of what he had been exposed to in Edinburgh. He returned a believer in evolution. The most famous stop on this five-year around the world voyage was the visit to the Galapagos Islands. But before the Beagle reached the western side of the South American continent, Charles Darwin had an experience at Tierra del Fuego that biographer Janet Browne said, "moved him most."[8]

The Beagle's Captain, Robert FitzRoy, had been to Tierra del Fuego on a previous voyage. This Land of Fire is at the southern tip of the

8 Janet Browne, *Charles Darwin Voyaging,* Alfred A. Knopf, New York, 1995, p. 250.

South American continent. While there, one of the small boats used to go ashore was stolen by some of the aboriginal people. In the ensuing chase of the thieves and recovery of the boat, FitzRoy found himself holding four hostages: two men, a boy, and a girl. He decided to bring them back to England with him and to experiment with converting and civilizing savage people. On return to England, one of the adults died of smallpox. The other three were given names: Fuegia Basket, Jemmy Button and York Minster. While in England they were taught English ways, acquired some fluency with the English language and were prepared to be returned to their native soil as missionaries to their own people.

When HMS Beagle sailed from England with Charles Darwin, the three Fuegians and the English missionary Richard Matthews were also aboard. FitzRoy planned to plant a colony of Christian civilization on the coast of Tierra del Fuego. This little colony would be the place from which Western civilization and Christianity would spread to the native peoples.

However odd all this might sound to our ears today, this sort of enterprise did not strike the English of the nineteenth century odd at all, including Charles Darwin. Englishmen of the nineteenth century viewed their world as the height of civilization and believed they had a Christian duty to spread its blessings wherever they could.

At the appropriate time the three civilized Fuegians and Richard Matthews were brought ashore and were met by their aboriginal countrymen. The little colony was established, and the Beagle sailed off on other errands. Nine days later when the ship returned, they found the experiment a failure. The little outpost had been attacked by the native people, the gardens trampled, supplies stolen, and Matthews was beginning to fear for his life. He was brought on board and the party sailed away leaving Fuegia Basket, Jemmy Button and York Minster with their aboriginal countryman. Darwin noted sadly that the three had returned to their primitive ways.

The importance of this episode for the theory of evolution was the effect it had on the thinking of Charles Darwin. He had sailed with the three Fuegians from England and had been in their company for some weeks. He had come to know each one as a distinct person. Before he met the Fuegians in their home environment, he was beginning to think that little separated someone like himself, an educated Victorian gentleman who had dined at high table at Cambridge, with those deprived of the influences of civilization. It was easy for him to believe at this point that, no matter the background, men were pretty much the same and that with the proper training and education, savages could be turned into gentlemen.

Darwin was rocked to his core by his first encounter with his shipmates' cousins. As the ship drew near to shore, he saw them for the first time amid the fires they lit on the banks (the Land of Fire) running and shouting wildly, unclothed with hair across their painted faces. He encountered an image that would figure in his thinking for the rest of his life. He wrote in his diary: "I shall never forget how savage and wild one group was…they were absolutely naked and with long streaming hair, springing from the ground and waving their arms around their heads, they sent forth most hideous yells. Their appearance was so strange, that it was scarcely like that of earthly inhabitants."[9] "How entire the difference between savage and civilized man is.—It is greater than between a wild and domesticated animals…I believe if the world was searched, no lower grade of man could be found."[10]

Janet Browne discusses the significance of this little-known episode in the first volume of her biography of Darwin. On board ship Darwin had lived with civilized savages. When he reached Tierra del Fuego, he saw these aboriginal people in their natural state. Browne states, "The ephemeral nature of civilization struck him most forcefully when he compared the three Anglicized Fuegians with members of their original tribes."[11] "Darwin's intense interest in the ability of human beings to change like this was fundamental to his later conversion to evolutionary theory… Without this experience in Tierra el Fuego—the experience of comparing Anglicized Fuegians with himself—he would never have had the breadth of vision to include mankind as an integral part of the natural world…"[12]

---

9 Browne, *Charles Darwin Voyaging*, p. 240.

10 Adrian Desmond and James Moore, *Darwin*, W. W. Norton & Co. New York, London, 1991, p.132–33.

11 Janet Browne, *Charles Darwin Voyaging*, Alfred A. Knopf, New York, 1995, p. 247.

12 Browne, *Charles Darwin Voyaging*, p. 249–50.

Browne concludes her discussion of this episode by stating that these events had a greater influence on Darwin's evolutionary thinking than anything else he ever experienced, including his work in geology and zoology. From his experience in the Land of Fire, Darwin drew the conclusion that civilization was but a veneer on the manners of man. He thought that aboriginal people were just a few steps from the apes and man was only a sophisticated animal. Man was not a different kind of animal, only the smartest of the animals. Nature was for him now all of a piece. From the lowest creature to the highest there was no longer any demarcation that set mankind off in a different category. Man was not a different kind of ape, only a smarter ape.

On board the *Beagle* Darwin would circumnavigate the world. He would be at sea for five years. Ahead of him was the experience of the earthquake in Conception, the Andes, the Galapagos Islands with its iguanas, tortoises and the now famous finches. He had with him the first volume of Lyell's book on geology. While he was on the western coast, the second volume would arrive by post. Lyell's view that the features of the earth had changed slowly and uniformly over a vast amount of time became Darwin's as well. For the rest of his life, he would be a firm believer in Lyell's uniformitarianism. As Darwin made his observations of the South American coasts, Lyell's view of geological change became his own. He began to apply Lyell's view of change to species as well. But of all that occurred during his five years at sea, Browne states that the experience with the civilized and uncivilized aboriginal peoples had the greatest impact on his subsequent theory.

In this abbreviated look at how Darwin came to believe that natural selection could account for the variety of species, it is worth noting that, like his experiences in Edinburgh, he is powerfully influenced, less by biological science than by events and situations outside of biology. He is deciding for evolution not by doing experiments in a laboratory, or by careful study of changes in species, but by inferences from his social experiences. The visceral man, not the rational man, is deciding for evolution. At Tierra del Fuego he witnessed the transformation of an uncivilized people to a state of civilization and then back to uncivilized. He drew the conclusion that men's behavior and manners were extremely fluid, and from this conclusion he went on to believe that biological forms are just as fluid. Biological forms, he would conclude, given millions of years, could change from one creature to another. In Darwin's day men of science believed that species were immutable, that is, species were, within limits, fixed and unchanging. Darwin was coming to believe just the opposite, that species are completely mutable, easily changed from one kind to another. He was moved to believe in evolution not because of empirical evidence, but because he saw Fuegia Basket, York Minster and Jemmy Button revert to their primitive way of life. But the question arises for the person reading Darwin's biography: what do civilized and uncivilized Fuegians have to do with biological change and the origin of species?

While Darwin was processing his experience in Tierra del Fuego with the failed attempt to turn uncivilized peoples into proper Englishmen, mulling over his geological findings on mountains and coral reefs, and puzzling over the variety of finches on the Galapagos Islands, his sisters were sending him books and letters from home. With their letters the sisters included pamphlets written by Harriet Martineau. Ms. Martineau was the rage of London society and the popular explainer of the ideas of the Reverend Thomas Malthus. Through his sister's efforts Darwin was being kept abreast of events and ideas in his home country. Back home, the hot political topic was reform of England's Poor Laws. The ideas of Malthus combined with Darwin's subsequent friendship with Ms. Martineau would exert yet another influence on his formation of the theory of evolution.

Reform of England's Poor Laws was the context for the social theory of Rev. Thomas Malthus. The poor received aid from the government, but aid to the poor seemed to exacerbate the problem and stimulated even more debate. How best to take care of the poor? How much could or should the government do to alleviate poverty? Whig reformers like Ms. Martineau used Malthusian theory to advocate a tougher approach to those depending on the government for help. Malthus's view of the plight of the poor and what to do about it was not a warm and compassionate one. His social theory stated that populations increase faster than the supply of food, thus producing a struggle for available food. In this struggle for food starvation is the natural result for those who cannot compete. Charity, as an answer to this problem, made the problem worse. By providing for the poor, it enabled them to continue to breed. Thus, the result of government charity and handouts were more poor people and an increasing need for charity to care for them. Charity, according to Malthus, did not solve, but exacerbated the problem. What to do?

The apparent solution, not surprising given the premises above: stop, or at least curtail the charity and send the debtors to prison or the workhouse. This is what was meant by the "reform" of the Poor Laws. Martineau wrote her pamphlets to sell this "reform" to the public: let the poor, on their own, compete in the struggle for existence and let the fittest among them survive. This plan would reduce the state's burden of expenditures and promised ongoing savings as the number of poor decreased in the Malthusian struggle for survival. While the experience with the Fuegians in South America had the greatest emotional impact on Darwin, the ideas of Thomas Malthus had the greatest intellectual impact on him. Darwin would be to his dying day a believer in Malthusian social theory and its application to the world of nature.

But these influences would not be fully felt until after his return home. As the Beagle neared Portsmouth, Desmond and Moore sum up what the five years on HMS Beagle meant to Darwin and to the future of science. He had seen so much and came home a changed man. He left England as an errand-boy naturalist. He was chosen for the voyage because Capt. FitzRoy needed a gentleman companion and because Darwin could also act as a collector of specimens for leading naturalists of England. Because of his work during the voyage, he was returning home as a prominent naturalist in his own right, with his own agenda, his own ideas, theories, and questions. He would no longer simply be at the service of other naturalists but would now theorize on his own and help chart the course of the natural sciences. He returned home with a head full of questions about geology and species and, most importantly, man. This is how authors Desmond and Moore describe Darwin's thinking as the Beagle neared home: though still early in his theorizing, Darwin was ready to see man as nothing more than a highly evolved ape.

As he sailed home, he had man in his sights. Having actually seen savage man he now questioned whether civilized men had descended from them. He could imagine that men like himself were the most highly evolved ape. "He had seen every grade in his human 'scale,' from the rude Fuegian—whose 'very signs and expressions are less intelligible to us than those of the domesticated animals'—through the Maoris, Tahitians, and pampas Indians, to the gauchos exterminating them. And all were being eclipsed by the British colonists marching under the Union Jack.…But how could this be? How could the same creator have made both the primitive and so sophisticated?"[13]

13 Adrian Desmond and James Moore, *Darwin,* W. W. Norton & Co., New York London, 1991, p. 190–91.

CHAPTER 3

INFLUENCES

This Whig restructuring was an assertion of middle-class Malthusian values. Darwin found that Malthus had acquired a new meaning… His doctrine of population, progress, and pauperism was no longer academic. It was the very kernel of poor-law policy: the stuff of inflammatory oratory, popular defiance, and government propaganda.

—From *Darwin,* by Adrian Desmond and James Moore.

Darwin returned to England in 1836 to find that the theories of Malthus had been implemented. Darwin would not read Malthus's famous essay for another two years, but he was coming home to a Malthusian world. The Reform Bill had been in place for four years. The poor were having to compete to survive or be sent to the workhouse. Industrial capitalism and competition were changing the England he knew before he sailed. The England he found on his return reflected Malthus's new rough view of the natural world. The leading lights of this changing England believed the natural world a dog-eat-dog affair and believed society should reflect this truth.

The old political order was being replaced by the new. The Tory/Anglican grip on government and society was being supplanted by dissenters, Unitarians and atheists. There was an ongoing revolution in industry and one in politics as well. As the Tories were losing governing power, Christian ideals were losing their place as the central governing precepts for society. Darwin returned to an England being transformed by the Industrial Revolution and the rough capitalistic ideas of growth and competition. Malthusian concepts were making for a leaner, meaner society that would have less truck with a sentimental faith. Darwin's Christian faith was never robust. He had always been attracted to any current anti-Christian ideas and was soon connected to those rejecting orthodox Christianity and enthralled with the latest social theory.

Home from the voyage, Darwin was torn between settling in the country or in London. He preferred the country but chose London in order to be near the naturalists who could help with the study of the specimens he had gathered. Thanks to his brother Erasmus, he would be in the inner circle of freethinking radicals. Erasmus was a close friend of Harriet Martineau, the popularizer of Malthus's ideas. Through Erasmus's circle of friends, Darwin would come to know not only Harriet Martineau but moderns like Charles Babbage. Darwin would be a part of a freethinking, smart set of London, which would have a strong influence on his developing ideas.

Charles Babbage was a mathematician and inventor. He is credited with inventing the first automatic digital computer: a hand-cranked calculator. Darwin met Babbage at one of Erasmus's soirees, and there experienced firsthand the way God's role in nature was being recast by the progressives of society. It had been common to believe that the Christian God regularly intervened in nature to produce new species. God as "miracle-monger" was the derisive term given to this view because it held that each new species required God's intervention. According to Darwin's new friends, this constant intervention made God appear both incompetent and capricious. The traditional view relied heavily on a literal interpretation of Genesis and a belief that the earth was approximately 6,000 years old. This view was based not on science, but on a not-so-subtle or perceptive view of the book of Genesis. It was an overdone creationism that allowed for no changes in species over time through any natural processes. This was also the view of the Oxford/Cambridge set which had held so much power for so long but were now seeing its long-held hegemony challenged by a different class of scientific thinkers.

According to Darwin's new friends, a different theory of origins was needed to show that God was not incompetent when he set creation in motion. Babbage, Martineau, Darwin's cousin John Herschel, the philologist Hensleigh would all help steer Darwin toward a different conception of origins. Sir Isaac Newton had shown how the world of motion was ruled by physical laws. God was a God of order. He was wise and capable of creating a world that was ruled by laws. His power was displayed in the regularity of these laws. Darwin's London friends wanted the biological world to be seen to work by principles similar to those Newton had laid out for the motion of objects. Creating new species miraculously did not fit at all with how they thought God should be doing things. But what if God created in such a way that, with the passage of time, the correct species appeared as if programmed? Babbage's hand-cranked calculator machine could churn out a certain sequence of numbers and be programmed to automatically change the sequence at a certain time. This rule-governed method is how Darwin's set saw God at work, if he was at work at all, in the creation of new species. Nature, having been set in motion, now rolled along, programmed to produce new species in machine-like fashion. This view allowed for God to be the originator of nature, but after that he was essentially out of the picture.

Babbage wrote an essay, *Ninth Bridgewater Treatise*, to answer a series of essays that trumpeted the view that nature showed God's design and purpose in created things. Babbage argued that a nature ruled by law was a greater compliment to the creator than a nature that needed his constant attention. His own hand-cranked calculator was again the metaphor for how nature was thought to work in the new natural science that was being formulated. In this new world that Darwin's set were helping to make, one no longer looked at nature through the eyes of a poet, but through the eyes of a nineteenth-century industrialist. Nature was not like a song but like a machine. She worked, not according to beauty and design expressing the character of a loving creator, but according to laws that were brutal and cruel like the society that the industrial capitalists were creating. The idea of nature running according to laws and competition was meat and drink for Charles Darwin. He would set for himself the goal of explaining nature in machine-like, law-like terms.

Just as Newton had shown it was not necessary for God to intervene to keep planets in their place, Darwin believed he was the man to show that God was not needed to bring new species into being. The new science would have little need of God. In fact, the new science needed God to be as far removed from the whole scheme as possible. Otherwise, you were not dealing with science, but myth and fantasy. Although Darwin eventually rejected Christianity and thought his theory explained adaptation and design minus any outside designer, he never completely removed some sort of god from his scheme of evolution. He still needed something greater than nature to jump start the whole show. A Christian he could not be, but by the end of his life he told people he preferred to be thought of as an agnostic rather than an atheist.

Newton had described the world of motion as a world governed by discernible laws. Darwin and his friends believed that the biological world should mirror Newton's world. Wild and chaotic nature would be seen so no more. She would be "brought to heal," and someone, Darwin, would show, like Babbage's computer showed, how regular and machine-like she really was. That the biological world mirrored the world of Newton, or that it degraded God to be involved in creating new species, or that species could easily change from one to another, none of these ideas are put to any kind of test or skepticism. Neither Darwin nor any of his friends recognized that the world of physics is strikingly different from the world of biology, or that the Newtonian model in physics might not be appropriate for understanding the complicated and messy world of biology.

Desmond and Moore tell us that "The issues were probably thrashed out during Eras's literary dinners. Here the sardonic Eras counterbalanced the serious Hensleigh, with the Unitarian necessitarian Martineau. Biology was not the only subject discussed by Darwin's friends. All were interested in German Biblical criticism and language studies and devoted to Whig Malthusian ideals. All were better read than Darwin, and he relished their conversation."[14] "Martineau's scientific attitude was typical of radical Unitarians. She saw nature as predictable, predetermined, invariant. It was subject to law and order, not the province of miracle." Nature was self-developing and "A self-developing nature

14 Browne, *Charles Darwin Voyaging*, p. 216.

held no terrors for them. Eras's group, with Martineau at its center, gave Charles the license to work out his own deterministic theories."[15]

That man's preeminence in the natural world was threatened by such thinking held no terrors for Charles Darwin. His friend Lyell shuddered to think that man was descended from apes. Such an idea degraded man and deprived him of his special place in creation. But Darwin had no such reaction because he had seen aboriginal man, wild and naked, in the Land of Fire. In the scale of living things he placed his own kind, Victorian gentleman at the top of the heap. Thus, Darwin's reaction to ape ancestry was not, "Oh no, I am descended from apes," but rather, "look how far we Victorian gentlemen have come from our ape ancestors, and from those poor souls in South America just barely removed from the ape."

Darwin had settled in London because he needed the help of the naturalists of his day. He had brought back a treasure trove of fossils, animals, birds and insects and needed the help of experts in the relevant fields to classify them. One of these experts was a young taxonomist named John Gould. Gould was to provide reasons to believe in transmutation rooted in natural science as opposed to German Biblical criticism or Whig political ideals.

Gould examined the finches that Darwin brought back from the Galapagos Islands. Darwin believed the birds represented varieties of the same species, but Gould believed that he was looking at birds that represented different species. If Gould was correct, then each island had its own species. Browne states that Darwin did not at first see the import of Gould's conclusions. But a few weeks later, Darwin heard Gould report on Darwin's South American "ostriches." Again, where Darwin thought he was dealing with varieties of one species, Gould believed the differences were enough to represent different species altogether.

Darwin now began to realize the significance of Gould's conclusions. He recalled how each island in the Galapagos chain seemed to

15 Browne, *Charles Darwin Voyaging*, p. 217.

have its own tortoise. He was beginning to see that because each island was somewhat different in geography and environment, those environmental factors resulted in different species. If Gould was correct, it not only confirmed Darwin's belief during the voyage that species could change but meant that living creatures were even more malleable than he had previously thought.

Examining the finches and tortoises, it appeared that each island environment could act on a descendant from a common ancestor to create new species. This insight regarding change with descent from a common ancestor was heightened by the fact that each island served to isolate the different ancestors. Browne states that Gould's work and his conviction that the differences in the finches amounted to different species marked a "turning point" in the life of Darwin.

The evidence from the different islands was scientific confirmation of what Darwin had long believed. From this point on, he had an unwavering commitment to transmutation, or as we would say today, evolution. Furthermore, he was now firmly committed to ape ancestry for mankind. His ideas about transmutation meant not just that different species descended from a common ancestor, but that man himself was only a highly developed ape.

If humans were only apes, then morals were simply part of the ephemeral trappings of civilization. Desmond and Moore tell of the continuing influence of Harriet Martineau. She was arguing against any "universal moral sense," claiming that among humans "right" and "wrong" are culturally conditioned, not spiritual endowments. Such moral relativism was common among radical dissenters, and Darwin was primed to appreciate it after his visit to Tierra Del Fuego and New Zealand. "Moral acts…had evolved from the social instincts to aid the cohesion of the human ancestral troop. Even the Christian precepts, 'do unto others as yourself' and 'love thy neighbor as thyself,' had evolved naturally out of our ancestors' 'sexual, parental & social instincts.' By the autumn of 1838, he had thrashed out all the social and moral issues with Hensleigh, Martineau, and Erasmus, and worked them into his evolutionary formula, enormously expanding its scope."[16]

---

16 Adrian Desmond and James Moore, *Darwin,* W. W. Norton & Co., New York London, 1991, p. 262.

As Darwin explored the world south of the equator on his five-year voyage, he had come to believe Charles Lyell's idea that small changes over a vast amount of time account for the geological features of the earth, and that this same view could be applied to the world of plants and animals. If great changes were to be accounted for in animals by very slight modifications, it was going to take a lot of time for changes of that kind to produce the variety of species now in existence. Lyell's books and the world as Darwin experienced it on the voyage convinced him that indeed vast amounts of time had passed and the vast amount of time needed for "slight modifications" to do their work was available.

Darwin began filling notebooks with his thoughts and questions about transmutation. His mind overflowed with speculations ranging over all sorts of related subjects. Browne notes that Darwin does not, at this point, require a "how." He is a man already convinced that transmutation is a fact and the "how" is not necessary. "…he possessed the pivotal idea of change in living beings—and of real ancestral links between animals and mankind. There was no need yet in his mind for an extra key—the key of natural selection—to lock it all together. What he had at the moment was gripping enough. All his personal ambitions, social background, and code of behaviour, and a wealth of half-hidden dislikes, desires, and shortcomings, coloured the rush of thought as much as any scientific reasoning."[17]

It is interesting to consider at this juncture how things might have been different had Darwin been able to settle in the country, perhaps close to Cambridge, as he so wanted to do. But he had to be in London because he needed the help of the other naturalists living there. We know that, upon his return from the voyage, he already believed in transmutation/evolution. He believed that species are not fixed, but plastic, with the ability to change from one to another. In London, only his friend Charles Lyell provided balance to Darwin's own thoughts on transmutation. Lyell had done much to start a revolution in science with his view that the earth's form was the result of very slow changes over vast amounts of time. Though he was a revolutionary as a scientist, Lyell was a traditionalist in his views of man, and as such counterbalanced the voice of Erasmus's radicals. Except for Lyell, Darwin's set were all of

---

17 Janet Browne, *Charles Darwin Voyaging,* Alfred A. Knopf, New York, 1995, p. 364.

one progressive mind, caught up in the capitalistic, competitive, dog-eat-dog world that English society was espousing and becoming.

But what might Darwin have heard if he were living closer to his old Cambridge friends? Those Anglicans who were now losing out to the new thinkers who saw nature as mean and machine-like might have provided some helpful caution. What if Darwin had his old Cambridge mentor Henslow to question him on all this radical thinking? Henslow could have provided a balance and a check to this rush to discard all the old orthodoxies. We will never know. We do know from Browne and Desmond and Moore that Erasmus's circle of friends had a great influence on Charles Darwin and thus his theory of evolution. And the theorist who had the greatest influence on him was the Rev. Thomas Malthus.

CHAPTER 4

MALTHUS

Almost imperceptibly,
he glided towards transmutation…

—From *Darwin,* by Adrian Desmond and James Moore.

DARWIN RETURNED TO ENGLAND AND witnessed a revolution in society and politics as entrepreneurial capitalists and Whig reformers changed English society. Moreover, he was intimate with some of the individuals responsible for those changes. While at sea, Darwin had read about the economic theory of Malthus that informed Whig political reformers and justified the government's harsher treatment of those in need. The reformers thought English society should be based on Whig ideals and capitalistic goals that correctly mirrored nature's own dog-eat-dog ways. Malthus provided the economic theory, which was then applied to politics, society, and soon biology.

Darwin's belief in transmutation reached a "turning point" with John Gould's assessment of the finches. Gould's insight had provided evidence for his view that species were mutable (not fixed). How these changes resulted in new species was still unknown, but Darwin's goal was to provide an explanation that did not require an intervening deity. At this time Darwin had a muddy sense of a God who was still somehow involved in the universe. He had never been a man really committed to Christianity and at this point in his life there was very little left of any belief in the Christian God. In 1851 Darwin would suffer a loss that would cement him as an unbeliever and thereafter forever hostile to Christianity. But before that blow, he found the key that had been missing from his thoughts on transmutation. The key did not come from biology, but it would make natural selection possible.

In 1838 Darwin finally read the *Essay on the Principles of Population*, by Thomas Robert Malthus. We have already noted that Malthus postulated that human populations increase faster (at a geometric rate) than food production (at an arithmetical rate). If there were no checks on population growth, said the Rev Malthus, people would outstrip the supply of food. But there were checks such as death and disease, late marriage, and abortion. These checks, according to Malthus, held the number of people in line with the available food supply. According to Browne, " …Malthus's inexorable numbers resolved his (Darwin's) evolutionary questions. …Darwin saw that what Malthus said about checks to fecundity in the human world rang emphatically true for animals and plants also."[18]

"Malthus further explained how this apparent balance (between population and food supply) emerged out of death—out of what Darwin … called the warring of species.'"[19] Darwin now saw this war as a creative force. Death of the unfit is a major component of Darwin's theory. Death came to the unfit in nature just as Malthusian theory brought death to the unfit in English society. Again, from Janet Browne, "Darwin's moment of insight came when he caught the idea that the ones who died would be the weakest and the ones who lived the strongest—or best adapted. Death, so to speak, could be a creative entity. It could preserve useful adaptations in some animals and plants while weeding out the remainder. Over a long period of time, favorable variations could become "fixed" in a population and a whole species might adapt to its current situation."[20]

And so Darwin, in his own famous words, had "a theory by which to work." The linchpin for the theory of evolution came to him from a social theorist whose ideas created the dreadful workhouses that Charles Dickens made so infamous, and whose ideas were used to make social policy in an increasingly capitalistic society. The ease with which he

18 Janet Browne, *Charles Darwin Voyaging,* Alfred A. Knopf, New York, 1995, p. 387.

19 Browne, *Charles Darwin Voyaging,* p. 388.

20 Browne, *Charles Darwin Voyaging*, p. 388.

adopted a theory that was not rooted in biology or the study of plants and animals is striking. Without hesitation, he applied Malthus's ideas about human population and food supply to plants and animals.

As we have seen, all of Darwin's time in London prepared him for an uncritical move in this direction. The influence of Erasmus and Harriet Martineau bore fruit when Malthus supplied the "linchpin" for natural selection and to Darwin's quest for a workable evolutionary theory. In addition, Darwin saw Malthusian ideas everywhere around him. It was easy for him to relate these to the natural world.

Job specialization in industry was seen in Malthusian terms. Darwin himself was part of the capitalist class that benefited from this "warring" that was the Industrial Revolution. He owned stock in various companies. Married to his cousin, Emma Wedgwood, he was related to the Wedgwood pottery empire. Darwin was well acquainted with the industrial and economic theories of the day that were making the Industrial Revolution possible. He understood how industrialists organized factories with distinct divisions of labor. One person was maximally efficient by doing one job repeatedly with each worker fulfilling a single simple function in production. He transposed the economic picture of mid-century England onto the world of plants and animals.

At the same time he was uncritically assigning Malthusian theory to the animal kingdom and enjoying the benefits of being a stockholder in a booming economy, England was at war with Russia in the Crimea. During the Crimean War, Desmond and Moore report that while Darwin was engaged in a last "major rethink of his theory, his thoughts were preoccupied with war and struggle." War, struggle, specialization of the labor force, death of the weak, victory for the strong, these were the factors that brought about progress. Darwin had been witness to such "progress" on his voyage, where he saw gauchos on the pampas dealing death to primitive peoples. Other peoples were enslaved by superior masters, when the British extended their empire in South America placing Englishmen at the top of the global scale. These thoughts, of war and the constant victory of the strong over the weak, formed the background for his last great "insight" into how he believed nature worked.

By the mid-nineteenth century, the picture of natural life as a branching tree was accepted by many naturalists, Darwin among them. But until this last "insight" he did not know how to explain it. He needed to explain how descendants branched off and were now different from their parents. Desmond and Moore describe that Darwin could remember the very spot, "whilst in my carriage," that the solution came to him. He believed he could now account for the forking of the tree of nature, i.e., he could account for the trunk, which was the common starting point or common ancestor, and its branches the different species.

"Just as his Malthusian insight had come from population theory, so his mechanism for creating diversity looked like a blueprint for industrial progress."[21] A discerning reader might wonder if Darwin would have reached his insights at all if he had not been living at the time of the Industrial Revolution. And it was a revolution. He lived on the cusp of this monumental event. In other words, this was definitely ***not*** the way things had been ordered for eons. He was ***not*** deriving an insight from a long and proven history of human conduct. This way of organizing labor, with factories and specialization, was brand new. There had not yet been time for wise heads to understand or come to terms with this new way of producing. This fact alone ought to have suggested caution for the author of *Origin.* Furthermore, there was nothing necessarily natural about this arrangement. How odd that Darwin, the naturalist, used a model that was in so many ways artificial and the product of man's greed and desire for efficiency. Though it was artificial, it was, importantly, the latest and most progressive way of managing labor. This new capitalistic industrial engine dominated his views and those of his contemporaries. The Industrial Revolution provided Darwin with the blueprint of how nature supposedly operated.

The Industrial Revolution changed the way people looked at work and workers. Men who for centuries had made a living by farming or by hand crafts were now herded into great central spaces. They worked for large business concerns whose primary goals had nothing to do with their welfare—monetarily, emotionally, or physically—but rather with the owner's pocketbook and the satisfaction of shareholders. Wealth was created and products were produced at less cost for the benefit of consumers. Jobs of a certain kind were created and the modern way of

---

21 Adrian Desmond and James Moore, *Darwin,* W. W. Norton & Co., New York London, 1991, p. 420.

life that we know today was being born. There were benefits from this revolution, but they did not come without costs. The worker in a factory had a job, but one that required him to trade his independence and pride in workmanship for a paycheck.

Whether it was worth the trade can be difficult to judge. For the man treated like a cog in a machine, the answer might be no. For a man like Darwin, who watched his income rise as his stocks paid dividends, the answer would be yes. The important thing to understand is that Darwin saw all of this as the *natural* order of things where the fittest prosper in the Malthusian struggle to survive. How could he not? He enjoyed the choicest position in the new order. The biological world of Malthus and the social world of Darwin were no longer distinct, but worlds blurred together. Would Darwin have reached his theory if he had not been born to wealth? How trustworthy can the theory of evolution be if it owes so much to the economic theories of the Industrial Revolution?

Again, from Desmond and Moore, "Darwin recognized that, just as industry expanded when the workers specialized, so did life ...He argued that natural selection would automatically increase the 'physiological division' of labour among animals caught in competitive situations. Just as a crowded metropolis like London could accommodate all manner of skilled trades...so species escaped the pressure by finding unoccupied niches in Nature's marketplace. The metaphoric extension was complete. Nature was a self-improving workshop, evolution the dynamic economy of life. The creation of wealth and the production of species obeyed similar laws. Division of labour was nature's way as well as man's."[22]

Desmond and Moore make one further telling observation. They note that when Darwin used the term "division of labour," he cited a zoologist's use of the term and not that of an economist. "To be successful, evolution had to be *seen* standing on the solid rock of science."[23] (emphasis mine). How telling that this theory must be *seen* to be standing on the solid rock of science, rather than the slippery rock of industrial economics. It was crucial that it should be so seen. Evolutionary ideas were

---

22 Desmond and Moore, *Darwin,* p. 420, 421.

23 Desmond and Moore, *Darwin,* p. 421.

entertained by many for most of Darwin's lifetime. What was lacking was a theory that was based, or at least appeared to be based, on science.

Darwin began his notebook on his species theory in 1838 but would not publish until 1859. In the interval between the beginnings of *Origin* and its publication came a book published anonymously. In 1844, Robert Chambers published a work entitled *Vestiges of the Natural History of Creation* or *Vestiges* for short.

Chamber's work proposed a law of development to explain all of nature. His work encompassed the development of nearly everything, beginning with the formation of planets from coalescing gases up to mankind. Life, the author said, was formed by some kind of spontaneous generation, and up and up things went until the law of development accounted for all the animals, current and extinct, including mankind. In 1844 the book was a publishing hit. Written by a largely self-educated, middle-class Scot who lacked the scientific pedigree of Darwin's circle and station, it was such a hit that many people who for years had ignored arguments for transmutation were now believers.

Darwin was stunned to see that someone else had a fully developed theory of evolution that was so much like his own. But *Vestiges* was not anchored in hard science of any kind, and it generated considerable criticism. Chambers's view was thoroughly materialistic and left God with little or no role in creation. Those who held to the traditional view of God's role in creation were especially vehement in their condemnation of Chamber's work. But the book paved the way for Darwin's own work. It showed him pitfalls to be avoided if he were to convince his own class of scientists of a thoroughly materialistic explanation of species. He decided that, when the time came, he must avoid any mention of the origin of life in his work and to leave that as a mystery unsolved. Above all, he must be seen to anchor his work in science.

To help anchor his theory, Darwin spent several years classifying barnacles, experimenting with pigeons and seeds, and spending hours gathering information from breeders of domestic animals. This scientific work would add to his knowledge of how species change on a small scale. But before discussing the importance of his work in those fields, it is well worth noting an experience that solidified Darwin's rejection of Christianity.

CHAPTER 5

ANNIE

After Annie's death, Charles set the Christian faith firmly behind him. Since writing his essay on evolution in 1844, he had held to the view he had reached then. Death was a purely natural process.

—From *Darwin, His Daughter, and Human Evolution,* by Randall Keynes.

For him the death marked an impasse and a new beginning. It put an end to three years of deliberations about the Christian meaning of mortality; it opened up a fresh vision of the tragic contingence of nature.

—From *Darwin,* by Adrian Desmond and James Moore.

DARWIN WAS WITNESS TO REVOLUTIONS in politics, economics, and religion. The Anglican Church, which had dominated English society into the nineteenth century, was losing its dominant position as it fell under attack from both dissenting Christians and non-Christians. Darwin was a part of this sea of change by his teenage years. As a young man he had considered becoming a country Anglican parson, but even at that age he could not assent fully to the Thirty-nine Articles. The Thirty-nine Articles were the essential statement of faith and doctrine for the Anglican Church.

As a teenager he was already uncomfortable with orthodox Christianity, which was not surprising given his family background. Desmond and Moore title the first chapter of their biography of Darwin "Catching a Falling Christian." They are quoting Darwin's grandfather, Erasmus, the author of *Zoonomia*, the work that extolled an evolutionary way of looking at the natural world absent a creating god. Old, unbelieving Erasmus applied this witticism to the Unitarian faith of Josiah Wedgwood. Since Unitarians deny the divinity of Jesus and the Trinity, he called Unitarianism "a featherbed to catch a falling Christian"—great phrase. The Unitarian faith, so stripped of Christian essentials, was not much more for him than a featherbed, and he did not think in the course of life the featherbed would do anyone much good. For Erasmus, it was much better to chuck the whole thing than to call yourself a believer of such an insipid faith.

It appears that Erasmus was right on the money when it came to his grandson Charles. The voyage on the Beagle helped erode what little Christian faith he had, and he now worshipped nature. On his return from the five-year voyage he was connected to people like Harriet Martineau through his brother Erasmus's social world where he was exposed to the latest in dissenting theological thinking. His drift toward an evolutionary view of life meant a further erosion of what remained of his belief in the Christian God.

For some, this is a nonissue. They do not see orthodox Christianity and Darwinian evolution as mutually exclusive. For some, God is in his heaven and has set in motion a creative process whereby the natural world is a product of what Charles Darwin described: descent with modification over time by natural selection.

On the other hand, there are others who do not see these two systems as compatible at all. They believe that Darwinian evolution and the Biblical account of creation represent competing views. Phillip Johnson has said that Darwinism or evolution is the naturalist's creation myth. It is, for those who accept it, the true story of our origins. It tells us how life began, where we came from, what we are made of, how we got to where we are now, where we are going, assuming we are going somewhere at all, and how all this will end. This alternative creation myth to the Genesis account of beginnings teaches that, among other things, life has no meaning, death is the end of life, free will is an illusion and there is no true basis for morality. That is, Darwinian evolution is more than just a way to view the origin of species. Noted evolutionist Richard Dawkins has said that Darwin's work made it possible for him to be an intellectually fulfilled atheist. Some may think Dawkins's view represents an extreme; but many, even if they have never consciously formulated it for themselves, live as Dawkins lives. Darwin's theory provides them with the basis for their worldview: a worldview that has a machine-like nature with no need of any meddling creator.

Whether reconciling evolution with Christianity is possible or even desirable, the important point for the story of evolutionary theory is how Darwin viewed the two systems. Darwin clearly saw that the more he came to believe in an evolutionary view of life, the less he could believe in the Christian God. There was no way for him, try as he might, to believe in both. This struggle over faith was especially painful for him because the class into which he was born—that provided him with so many friends and teachers—was made up largely of believers. Even more difficult was the fact that his wife, Emma, whom he loved dearly and who loved him, was a believer.

Emma Wedgwood was brought up in her parents' Unitarian faith. She was, in spite of this, a believer in Jesus and his promise of everlasting life. She read her Bible and found the Gospel of John particularly edifying and important. Early on in their marriage, she understood that her husband's work was moving him further and further from Christianity. Emma believed she needed to speak to him about the implications for him and for them. In short, she feared for his soul. She painfully considered the possibility that he would not be reunited with her in heaven after their deaths. Believing that John's Gospel conveyed the greatest spiritual truths most beautifully and convincingly, she urged Charles to consider carefully the farewell discourse that occurs near the end of the John's work. In this discourse Jesus uses the picture of himself as a vine and his followers as branches. It is in this passage that he issues the famous new commandment to them, to love one another.

Charles understood his wife's motive and did as she asked. He was moved to tears by her honesty and concern and wished that he could reassure her that yes, he could believe in this. But he could not. As much as he wanted to grant Emma the answer she wanted to hear, he could not. What Emma hoped to be some kind of breakthrough in faith for her husband became an occasion for tears instead. For Charles Darwin, it was not possible to believe as Emma did and believe what he believed to be true about the natural world. But the clincher for Darwin personally was yet to come. Ahead in their family life was an event in 1851 that would cement Charles's unbelief.

In the years following Darwin's reading of Malthus's essay, Darwin became a voracious reader. He and Emma read biographies, philosophy, social theory and religion. At this same time Darwin's father's health was failing, and Charles was experiencing vomiting, dizziness and headaches. He spent hours working on his tiny barnacles, inhaling the fumes of preservatives for hours; it was tedious work with a primitive microscope. But Darwin was also strained by his growing conviction that his evolutionary view of life was true. He had no illusions about the implications of his theory. Belief in the immortality of the soul, the basis for morality, the belief in free will—all these would be in tatters because his theory meant there was no supernatural being who superintended the universe. Nature for him was becoming the whole show. As he grew in his conviction that evolution was true, he felt more and more a traitor to his class. The gulf between him and Emma grew.

His father's health continued to fail. His own health sometimes made him feel that he would not live long. Plagued by the implications of his work, he continued to see if he could find his way to some sort of belief. The fruits of German Biblical criticism had taken hold in England. It was easy to find authors who had abandoned orthodox Christianity because German theology made it seem naïve and out-of-date. Darwin was attracted to the German sophisticated skepticism and dismissal of orthodoxy. At the same time, many modern authors were trying to maintain some kind of faith after dismissing an historical, divine Jesus, and to reconcile an unhistorical Jesus with modern science.

Desmond and Moore tell us Emma had Charles read Andrew Norton's *The Genuineness of the Gospels*, to give him reasons to believe in eternal life. But they note that he soon turned to reading the story of a fallen curate, John Sterling. He was critical of Sterling's "emotional attachment to Christianity." They quote Darwin telling his clergyman-cousin Fox, "'I simply cannot believe in the same spirit…that ladies do believe on all & every subject.' Maybe he could no longer believe at all."[24]

Following the story of Sterling's spiritual struggle, Darwin turned to Coleridge. Darwin read his *The Friend and Aids to Reflection.* Perhaps here Darwin would find a man who could help him find his way to reconcile some sort of faith with science. But Coleridge's view of religion "required a leap of faith that Darwin could not make. For Coleridge, religious feeling was innate to the soul and had nothing to do with inherited instinct, but Darwin's heretical notebooks belied this. Coleridge was wrong."[25]

A complicating layer to this struggle of science and faith was the Christian idea of hell. Contemplating his father's impending death, it was natural for Darwin to be wrestling with the Christian idea of eternal punishment. The doctrine of hell is troubling for anyone, no matter what their religious convictions or thinking. Darwin had always found the Christian teaching on hell difficult and now that he was facing his father's death the difficulty was magnified. The thought that unbelievers suffered eternally was for Darwin "monstrous," according to Desmond and Moore. His father was an unbeliever and so was Charles. Hell being monstrous, he found no comfort in Coleridge's view that a lack of feeling for Christianity resulted from an enslaved will. Darwin's father died in 1848 and Charles grieved like any son would but did not experience any personal crisis as a result of the loss. He simply rejected the idea that his father could be in a state of eternal punishment.

---

24 Adrian Desmond and James Moore, *Darwin*, W. W. Norton & Co. New York London, 1991, p. 360.

25 Desmond and Moore, *Darwin*, p. 360.

In 1849, his friend Harriet Martineau published *Eastern Life Present and Past*. Not surprisingly, Darwin was sympathetic to her view that "Christian beliefs about reward and punishment were based on heathen superstitions."[26] His reading of authors like Norton had convinced him that Christianity had to rest on a solid basis of evidence and not the vague spirituality of people like Coleridge. But he did not find the evidence persuasive enough. He was uncovering evidence to the contrary that for him was much more compelling.

By 1850 Charles and Emma had a family of seven children. Nine-year old Annie was her father's favorite, and she was ill. Charles feared she was ill because she had inherited his own weak constitution. In the coming months, as Charles and Emma did all they could, and all that nineteenth-century medicine could do to make her well, Annie got progressively worse. Two years after the death of his father and while struggling with the frightful possibilities of Annie's failing health, Charles was reading Francis Newman.

He had been able to handle the death of his father. But now he was faced with the possible loss of his favorite child. Though no Christian, he wanted to maintain some sort of belief, wanted to find some halfway place for religion and science. He was walking a fine line and it was natural that he continued to read. He was searching for some sort of theology from the latest spiritual writings of the day that would bring him comfort. But the latest spiritual writings would be of no help, rather they made things worse. German Biblical scholarship was creating skeptics by undermining the trustworthiness of the Bible, the historicity of Jesus, and any belief in miracles or the supernatural. The religious lights of Darwin's day were mapping out a vague spiritual world where one is somehow connected to God, but without dealing with terms like sin, redemption, the cross and sacrifice. These moderns asked the reader to forget the hard old doctrines and to look to an updated version of God. This God looked each year less and less like the one found between Genesis and Revelation.

Desmond and Moore tell us as Darwin faced his greatest personal crisis and loss, he was reading Newman, a former Latin professor at University College and brother of Cardinal Newman. Even among

26 Desmond and Moore, *Darwin*, p. 362.

Unitarians, Desmond and Moore note Newman was regarded as a "free-thinker." Thanks to Robert Chambers's *Vestiges,* Newman was already an evolutionist. Having abandoned basic Christian doctrine, he endeavored in his writing to achieve a workable synthesis between science and religion. "He rejected the 'dreadful doctrine of Eternal Hell.' Real assurance of a blessed hereafter, he maintained, came only through 'a full sympathy of our spirit with God's spirit.'"[27]

This sort of thing might have been right up Darwin's alley, but at this time in his life he needed much more. It is one thing to find a religious alternative to orthodox faith satisfying when one is comfortably resting with one's book and enjoying the thought of being part of the latest debunking that your circle of friends do so well. It is quite another to discover how comforting modern religion can be when the hard chips of life are really down. Newman was Darwin's cup of tea, but he was already farther down the road of unbelief than Newman was. Even though he was fully sympathetic to Newman's religious view, he found it hard to believe in this most watered-down faith.

He read *History of the Hebrew Monarchy*, but it was no help. Even though it followed the latest German scholarship, Darwin noted that "the book left 'the relations between the divine and the human mind...the same as ever.'"[28] Darwin wondered, "How could faith in the Holiness of God arisen amid the death, famine, and wars of Semitic tribes? No, Charles insisted, the religious instinct had evolved with society. The primitive Jewish God, whose atrocities had lit up hell fires in Christendom, could be nothing but a barbaric tyrant. Newman was consolidating Charles's doubts."[29]

Annie was not getting better.

He read Newman's *Phases of Faith* and found echoes of his own life in Newman's pilgrimage to the outer edges of belief. But Charles could not stop and rest where Newman did. He could not live with some mishmash of science and religion. He needed more. There had to be for him real evidence if he were going to believe. Vague talk about "a full sympathy of our spirit with God's spirit" was not what a man like him

27 Desmond and Moore, *Darwin*, p. 376.

28 Desmond and Moore, *Darwin*, p. 377.

29 Desmond and Moore, *Darwin*, p. 377.

could feed on. Newman's halfway spiritual house of belief and science would not do.

"There was no stopping short at Unitarianism, that 'featherbed to catch a falling Christian,' as Charles's grandfather said. Christianity had to be rejected once and for all. The moral logic that condemned eternal punishment could not condone the New Testament where the monstrous doctrine was taught. Evolution—the new gospel—explained mind, morality, and religious beliefs as part of the social development of the human species.'"[30]

For relief from his physical ailments Darwin had sought help at different times from a Dr. Gully who prescribed water cures for Charles and had a clinic in the Malvern Hills. Annie was now in greater need of Dr. Gully's treatments than Charles. In March 1851, Charles left Emma at home with the other children and took Annie and her nurse to Gully's clinic in hope of a cure. Annie would battle her illness for a month with Charles living with the ups and downs of her improving then declining health. But Gully's medicine was no match for her ailment, and she died April 23, 1851. The loss devastated Darwin.

From this point on Charles Darwin would have no interest in anyone's attempt to reconcile some sort of belief in God with what he believed to be true about the natural world. Desmond and Moore conclude this episode in his life with these words: "This was the end of the road, the crucifixion of his hopes. He could not believe the way Emma believed—nor what she believed. There was no straw to clutch, no promised resurrection. Christian faith was futile."[31] "Annie's cruel death destroyed Charles's tatters of belief in a moral and just universe. Later he would say that this period chimed the final death-knell for his Christianity, even if it had been a long, drawn-out process of decay. Charles now took his stand as an unbeliever."[32]

In 1851, Charles Darwin experienced the kind of loss that was not uncommon given the times in which he lived. But though the loss was common for nineteenth-century families, the result was not. Fathers must cope as best they can with such pain whether it is the common

30 Desmond and Moore, *Darwin*, p. 378.

31 Desmond and Moore, *Darwin*, p. 384.

32 Desmond and Moore, *Darwin*, p. 387.

lot or not. For Darwin the loss was a blow from which he was hardly able to recover and which marked him for life. We can only grant him much sympathy and understanding. But any chance that some residual Christian faith might leaven his program to construct a view of nature where there was no God was now gone.

Since his boyhood when he was so engrossed in the world of beetles, he had a view of nature that was particularly dark. Observing their "struggle" for life and seeing nature wasteful and chaotic, he saw nature at "war" and her creatures in near constant pain and agony. The possibility that insects do not experience physical pain and mental anguish as humans do seems to never have crossed his mind. His life's thinking was one of linking man in smooth continuity with nature. To some degree, in Darwin's developing worldview, Annie was simply another creature that failed to survive because she was unfit. The "war" had reached him personally. Her death was just another in untold numbers of deaths that occur in nature every day. Her death was a natural process that did not involve God or any moral issues. The death of the weak was the way nature worked. Tomorrow Emma or he could be felled from the same ongoing struggle for survival.

But such a view of life cannot be the whole story—Annie Darwin's death the same as the death of a beetle! Even as he believed this about her death and its place in nature, his heart was broken. Of course, it was broken because he and Annie were not just creatures like beetles involved in a struggle for survival, but human beings. Unlike a beetle, Charles had a heart, a heart that loved his little girl. The heartbreak was real, as real as any datum of evidence he had collected all his life. But his view of nature would not allow for the truth of a broken heart to show him where his thinking could be wrong. This horrible pain was not enough to get him to question whether in fact he and Annie were nothing more than creatures caught up in a struggle of the survival of the fittest. He had convinced himself that natural selection had not only produced all the animals and plants we know but had somehow been able to produce a heart that can be broken by the death of a dearly loved child. The pain that might have awakened him to the truth that he was more than just an intelligent ape was not enough to override his belief in natural selection. For Charles Darwin, nature operated by laws and those laws applied equally to people and bugs and cared not a bit for either.

As he sought to find his way during this time of impending loss by reading the latest books on spirituality, he was singularly unfortunate in his choices of guides. The authors that we have mentioned did him more harm than good. Desmond and Moore in *Darwin* and Randall Keyes in his book, *Darwin, His Daughter, and Human Evolution*, both single out an idea from Newman's *Phases of Faith* that must have struck Darwin particularly hard as he contemplated Annie's suffering. In this work, Newman told the story of his own journey into unbelief. It was a journey that mirrored Darwin's own as Newman found "element after element" of Christianity wanting.

For Newman, as for Darwin, the idea of hell was a supreme difficulty. Keyes notes, "Newman argued that the Christian doctrine of eternal punishment had no clear scriptural authority and suggested that it posed intolerable moral difficulties. One such difficulty was that by the reasoning of some theologians every sin was 'infinite in ill-desert and in result, because it is committed against an infinite Being.' 'Thus, some said, 'the fretfulness of a child is an infinite evil.' "[33] There are numerous problems thinking about sin in this way, but what Darwin took from it was that God is ready to send a ten-year old child to hell for eternity because of her fretfulness. Whatever remained of Darwin's Christian faith in 1851 was snuffed out with Annie's death.

What does Darwin's loss of faith have to do with the theory of evolution? Part of the answer involves correcting in the minds of many the notion that Darwin arrived at his theory in complete objectivity. Many hold the notion that Darwin represents the pure pursuit of scientific truth. He is often popularly portrayed as one free from the prejudice of his day: a scientist who freed us all from the superstition that is faith. However, modern biographies place Darwin in the context of Victorian England. We learn that the social forces around him, the "spirit of the age," and the death of his daughter had an enormous impact on his thinking. We know there were powerful nonscientific factors that went into *Origin.* It is certainly true that, in the mid-nineteenth century, the picture of

33 Randall Keyes, *Darwin, His Daughter, and Human Evolution,* Riverhead Books, New York, 2001, p. 176.

the origin of species was in need of correction. In Darwin's day, it was thought that, because God created in Genesis, there was no room for any kind of change of species unless he directly intervened. This represented a form of creationism that failed both as science and sound theology. There was a need for a rethinking of how species changed over time.

But in correcting a view that was all Bible and no science, Darwin aimed not to correct but to revolutionize our view of nature. He reached his world-shaking theory by riding the radical currents of thought in his day as much as by scientific experiment. His rejection of Christianity meant that he was now free to remove God, and God's power to intervene in nature from his evolutionary equation. He did not, however, eliminate God completely from his scheme, but the god that is left after his theory is superfluous. Beginning in 1838, as he worked in his notebooks, Darwin found as his theory grew, what little Christianity he had was dying. His wife, Emma, saw this clearly and tried to get him to consider the truths that she believed. She understood that Christianity would provide a kind of brake on a thoroughly naturalistic scheme of origins. But even before Annie's death Darwin could not consider the Christian ideas Emma encouraged.

When he published *Origin,* he avoided the question of the real origin of all things as something so far back in time and so enveloped in mystery, that the question was unanswerable. But with Christianity rejected and the absolute origin of all things unsolvable, one could still call on some sort of god to get the whole show going. As an agnostic he was able to allow for an unknowable and uninvolved deity that could be beyond scientific inquiry. Such a god was useful for explaining the very beginning of the universe in such a way that once the show is up and running, he could be dismissed.

From the Industrial Revolution Darwin drew the lessons of never-ending competition and adaptation. From Chamber's *Vestiges* he accepted the tree of life as a picture of how species changed over time. This picture meant for him that all life came from one or a few common ancestors. Later generations changed because natural selection enabled them to fit into a particular niche, just as factory workers fill a particular niche in a factory.

In addition, he was caught up in the skeptical rationalism of friends like Harriet Martineau. German Biblical scholarship that cast doubt on the trustworthiness of the Bible and the historicity of Jesus was the order of the day along with the debunking of the old orthodoxies. All his friends were caught up in Whig idealism and political reforms and above all the social theory of Thomas Malthus. Darwin never looked back after reading Malthus in 1838. Add to those factors his total rejection of Christianity. Christianity was the one worldview that stood in the way of a completely natural explanation for every development and change in species over time. Thus, we come to an understanding of the origins of *Origin* that places it in Darwin's social context. It was hardly just the product of scientific investigation and experiment.

It is evident that what we see is greatly determined by what we believe, and scientists and intellectuals are no exception. This is particularly true of Charles Darwin, who from his early years saw nature not as a world of beauty and poetry, but of struggle and death. From his youth he was focused on beetles not eagles. The naturalists who preceded Darwin looked on the same natural world that he did. They looked at the same species that he did. Many saw beauty, purpose, and design. Darwin saw waste and cruelty. Many saw the hand of God. Darwin thought the idea of a loving god ridiculous in light of nature's cruelties. Many of his contemporaries had considered the idea of transmutation, evolutionary change in species for years and rejected it because of the lack of evidence, yet Darwin saw it differently.

What was the evidence he presented to convince the Victorians that their view of the created order with purpose and design was mistaken? We now turn to that evidence and the events that followed the publication of *The Origin of Species by Means of Natural Selection or The Preservation of Favoured Races in the Struggle for Life* to learn why Darwin's theory carried the day with so many of the intellectual class of his time.

CHAPTER 6

BARNACLES AND PIGEONS
ANALOGY AND EXTRAPOLATION

Through the sheer power of imagination he saw how species could be joined together by descent—a story of development linking geological antiquity with the present day.

—From *Charles Darwin Voyaging,* by Janet Browne.

From what we learn of Darwin's life before he embarked on his five-year voyage, we know he was primed to believe in transmutation. It was an idea that had gained the acceptance of many, but it needed grounding in science to be accepted by the more critical circle that was Darwin's educated class. During the voyage he would have experiences that would do much to confirm what he was already thinking. What he experienced a world away from England confirmed what he believed or nearly believed before he left England. He returned and provided the scientific grounding that would make the theory of evolution acceptable to his fellow naturalists.

Darwin voyaged around the South American continent and the world reading Charles Lyell's works on geology as he sailed. Lyell's belief that changes occur slowly and by the same forces seen at work today would greatly influence a generation of natural scientists. Charles Darwin, changed by Lyell's work, would be the most influential of that generation. The world was much older than many had previously thought. With enough time, no matter how small the changes being made, it seemed anything might be possible. Darwin accepted Lyell's view and never looked back. *Natura non facit saltum*, nature makes no jumps, would be forever a part of Darwin's gospel no matter how difficult it was to demonstrate or how lacking the evidence. Darwin was sure he could apply Lyell's way of thinking to the world of living things. Given enough time, small changes could account for all the different species.

The most significant experience he had during the voyage was with the aboriginal Fuegians and the civilized Fuegians. The impact was immediate, allowing him to connect man with nature, indeed to connect all living creatures in one system. Man was no longer for him a creature of any special kind, but simply the most highly developed creature that had descended from lower apes. His most significant experience with animals would be with the finches on the Galapagos Islands. But he would not see the significance until his friend Gould showed Darwin that what he thought were varieties of finches were perhaps different species of finches.

He returned home a believer in transmutation or evolution, but in need of a mechanism to explain how evolution happened. He was helped by his circle of friends in London. They encouraged him in his revolutionary thinking and in debunking the current orthodoxies of his day. During the twenty years of compiling his notes, he was caught up in the Industrial Revolution and the prosperity afforded him as a venture capitalist. All signs of Victorian society and Great Britain's expanding global dominance confirmed what was already in his head and heart. All of nature was at war and the "favoured races" were the ones who survive. But he was still without a mechanism to explain how all came to be. What could account for what he believed to be true?

Then, in 1838, came his breakthrough via Malthus and he now had "a theory by which to work." But he sat on his theory for years, fearing to share with the world what he believed to be true. He knew what his theory would mean for the Christian world in which he lived. Among its many repercussions would be the end of meaning for the universe and the end for the traditional basis for morality. No wonder he was reluctant to share.

What jolted the reluctant naturalist to finally publish was a blow that came from halfway around the world. Alfred Russel Wallace, a collector of specimens who was living in the Malay Archipelago, sent to Darwin his own theory of evolution. Darwin was stunned. In his own words he had been "forestalled." In all the twenty years of keeping his notebooks and keeping his counsel, it never occurred to him that someone else could arrive at so similar a theory. Their theories had differences, but they essentially mapped out the same scheme to account for the origin of species. He would now have to go public and share this idea with another. On July 1, 1859, with Darwin at home and Wallace a world away, a paper was read from each man at the Linnean Society. Neither paper caused much of a stir that night, but Darwin now set to work to put his ideas in book form. After twenty years of thinking and dithering he wrote *The Origin of Species by Means of Natural Selection or the Preservation of the Favoured Races in the Struggle for Life.*

We have considered many of the influences in Darwin's life that made him a believer in transmutation. We also need to consider the science that went into the theory. We cannot in the limited space here cover all of the experiments he carried out over the course of twenty years, but let us look at two that had particularly deep significance for him. During the years of recording his thinking in his notebooks, 1838–1859, he worked with barnacles and pigeons. The work with barnacles was triggered in part by a remark from Darwin's good friend Joseph Hooker. At a time when Darwin was filled with confidence as a general naturalist to write about origins, Hooker let him know that he did not believe anyone had that right unless he had studied at least one species in detail. The criticism hit home. Darwin certainly possessed an enormous amount of knowledge of geology, animals, plants and nearly all things natural, but he quickly saw the merit of his friend's challenge. How could he speak about how species change over time if he had never studied even one of them up close? He set about to remedy his situation by examining a species of barnacle. He thought this project would take about a year. It took eight. It became one of Darwin's great scientific labors.

He began in 1846 working with a creature that was little understood in his day. His was a pioneer's work in classification. From his close examination of the tiny creature using his microscope he would be able to show that what appeared to be a parasite was actually the male barnacle attached to the female barnacle. In some cases, the female had not one husband attached but two. Darwin was fascinated by this odd sexual arrangement. He believed it showed his evolution theory was correct because, he reasoned, God could not be a party to such an arrangement in the animal kingdom.

More importantly, Darwin theorized before ever working on barnacles that the male/female sexual arrangement in nature was something that had evolved from species that were once hermaphrodites. While some barnacles were hermaphrodites plain and simple, with the species having both male and female parts, another species possessed semi-hermaphroditism with what Darwin termed "complemental" males. Finally, he discovered species that had both males and females.

His discovery of the barnacle's odd method of reproduction did not confirm his theory from just biological reasons, but for emotional reasons. Darwin did not believe God would be a party to a universe in which there was such an odd reproductive arrangement even for a creature as lowly as the barnacle. Viewing the barnacle's method of reproduction, he was repulsed. Since God would not be a party to such an arrangement, then this must be the work of godless nature. He felt his theory confirmed.

He also concluded from his barnacle studies that the different species of finches that came from a common ancestor on the Galapagos Islands were not the exception but the rule. Up to this point he figured that the change from specie to specie was a rare occurrence. But studying the tiny barnacle with its strange sexual ways turned him 180 degrees. Because his barnacles varied in so many ways, he now thought that changes were always ongoing, that species were easily changed by natural selection and therefore classification schemes were arbitrary. If species were nearly always in a state of becoming something else, then specie was not a fixed category but an arbitrary classification with little meaning. Evolution was going on all the time.

His other major experiments were conducted with pigeons. Pigeon fanciers of his day were often part of clubs that shared knowledge of breeding pigeons for particular traits. This was domestic breeding along the lines of what men did with cattle, sheep, and dogs. Darwin got into the act. He had his own aviary built. With his own birds he found he could effect changes generation to generation. Here was transmutation right before his eyes—evolution in action without a doubt! He was able to manipulate certain features such as the crop or beak. The transformation that Darwin intuited at the Galapagos Islands was visible up close at his home in Kent. His theory of evolution would therefore be, in one sense, a grand analogy. He could speak to a fact that his readers acknowledged: domestic breeders—whether with cattle, dogs, sheep, or pigeons—can guide changes in the animals in their care over generations to produce a variety or develop a specific characteristic. According to his analogy, what the English farmer did with his cattle, and what Darwin did with his own pigeons, nature did in the wild. Men guided artificial selection, breeding for certain results or effects, and nature did a similar thing on a grander scale. All we need add to this equation is enormous amounts of time for nature to make the necessary changes. Given enough time, all that was then known in nature could come from one or a few common ancestors. Darwin figured he needed 300,000,000 years.

From the lessons he drew from Malthus and the Industrial Revolution, he viewed species as creatures under pressure to survive and find their proper niche in their environment. Animals were like factory workers finding the one simple job they could do day in day out to produce the factory's product. Because of the pressure on creatures to survive and find that niche where they could thrive, natural selection, by saving the helpful variations, continually changed creatures until they had the forms we see today.

One cannot read Darwin's great work and not be impressed. His knowledge of the natural world was astonishing. Not only had he experienced much and studied much, but he spent years in correspondence with other naturalists from all over the world. He cites examples from Australia, South America, the South Sea Islands, and North America. Telling points were at his fingertips. On and on he could go, piling up example and fact one upon the other so that he nearly overwhelms the reader with details. A gracious and gentle guide for the most part, he candidly dealt with objections that he knew the reader must be contemplating. He knew he was asking the reader to swallow a great deal and the reader had his sympathy. After stating a possible objection, he did all that he could to show why natural selection did what he believes it had done. He was seldom dogmatic, and the words he used, for the most part, show an understandable lack of assertion.

Given that many already believed in evolution and many more were disposed to believe it, the breakthrough that Darwin was to share with the world was the "how" of natural selection. We know that the idea of evolution or transmutation, the term used in Darwin's day, was already accepted by some and had been popular for decades, but no one had yet shown how it might have happened. There was not yet on the table a plausible scientific means to account for species evolving one from another. Natural selection was Darwin's great insight. It provided the science of his day with a scientific explanation for evolution. At the heart of the theory lies an analogy. Man working with domestic breeds selects for certain features in a process known as artificial selection and so what nature selects unaided by man Darwin calls natural selection. He believed there was no limit to the degree of change a species could undergo if natural selection had enough time. He believed there was enough time, and therefore the changes required were possible. With natural selection, 300,000,000 years, and "the sheer power of his imagination," he could explain all species as descendants from one or a few common ancestors.

Of all the claims made for natural selection in *Origin* the one that can be most readily accepted is that natural selection confers on an animal an advantage for survival. (By the seventh edition of *Origin* he was using Herbert Spencer's famous phrase: the survival of the fittest. We realize this famous phrase is not saying much at all. Of course the fittest survive, that is how we know they are fitter than their competition. Would anyone propose the survival of the weakest?) But nevertheless, one can imagine that forces in the environment working on a species could affect beneficial changes that could be passed on to succeeding generations. The changes would help the animal to survive, and the advantage would be passed on to subsequent generations. And, as with the case of artificial selection, over time an aspect of the creature could be altered. To cite a famous example, the lengths of the beaks of Darwin's finches on the Galapagos Islands have been shown to have increased or decreased in response to changes in their environment. The change in length helps the bird to eat seeds, the environment now favoring one length, later another.

The beak of the finch undergoes evolution in response to environmental pressures. It is perfectly reasonable to see how certain traits or characteristics would change for a species over time. A creature's legs might grow longer or stronger over time, or it might become a faster animal, or any number of other slight variations.

But if that were all that Darwin was proposing his book would not have sold out in its first day. He was proposing much more. If all he claimed for natural selection was that it helped a creature survive, who today would know the name of Charles Darwin? The fact is Darwin's name is as recognizable as Shakespeare's or Beethoven's. For many it is a name that means scientist. Darwin is famous because from the analogy of domestic selection he makes a breathtaking extrapolation. This mechanism of natural selection that confers an advantage for survival is so powerful, and species are so pliable, that we not only see the fittest survive, but derive new species in the process. Wings, lungs, eyes are produced. Reptiles become birds, whales leave the land and become creatures of the sea. Through small changes over vast amounts of time, one species begets another. This is the significance of the word origin in the title *Origin of Species*. Science in Darwin's time held that species were fixed, that is they did not change or morph from one to another. The term used in that day was that species were immutable. Darwin said otherwise. We can observe small changes within species; by a process of extrapolation, we can imagine foundational changes. The origin of a creature is the creature that preceded it in time, and so on and so on back to the beginning of the first creature.

From a well-established naturalist, not a popularizer of the penny press, this was an extravagant claim. *Origin* was no mean little book, though its author worked to keep it as short as he could. The writing is skillful and engaging. But just how strong a case was made for this claim? In all the wealth of words, what is it that persuades the reader that Darwin is right? Just how strong is the science behind the theory? We will review what some of Darwin's most astute critics said after publication of *Origin*. But before we consider some of the criticism leveled at the theory in Darwin's own time, let us take a moment to consider why science believed species were fixed in the first place. Let's ask why most scientists of Darwin's day thought otherwise.

CHAPTER 7
THE CRITICS

Is it necessarily good that a sponge or a polyp should be promoted to the rank of an oyster…?

—William Hopkins

THE IDEA OF EVOLUTION HAD been batted around thinking circles since Darwin's youth, and by mid-century many people believed in evolution, but the leading naturalists of the day did not. It is well worth asking ourselves why. Why did Swedish botanist Carl Linnaeus, founder of modern taxonomy; French biologist Georges Cuvier, expert vertebrate paleontologist and comparative anatomist; Richard Owen, Britain's leading fossil expert and founder of the term dinosaur; and Louis Agassiz, Swiss/American naturalist and discoverer of ice ages, all believe otherwise? If the evidence for all species being the result of descent with modification over time from a few common ancestors was so strong, why did the leading experts in the various fields not see it? Or, what *did* they see in their study of species, anatomy, and fossils that led them all to believe the opposite. While it was acknowledged that species varied within type, none believed or saw evidence that one species led to another.

Our present system of classifying species as we do—i.e., Kingdom: Animalia, then Class: Vertebrata, then Order: Primates, then Genus: Homo, then Species: Sapien, was the result of Carl Linnaeus's (1707–1778) study of nature. His foundational work in showing how creatures are related is familiar to us all, and it was based on his close study of nature. In spite of decades of dominance of Darwinian evolution, we still use his classification system. The fact that Linnaeus's work has survived in spite of the dominance of Darwinian evolution is a testament to the enduring quality of this pioneer's insights. No matter that nearly all of us believe in some kind of Darwinian evolution, we still go to the zoo to see different kinds of animals. The birds are found in the aviary, the large cats are in open displays, the reptiles in the herpetarium. We can confirm Linnaeus's work every time we go to the zoo. There are no displays of transitional creatures.

Richard Owen (1804–1892) created a classification system for fossils. He was the leading taxonomist in Great Britain and superintendent of the British Museum. At one time he was a friend of Darwin, but the friendship turned sour after the publication of *Origin*. Georges Cuvier (1769–1832), the great French paleontologist and anatomist, showed that fossils came from formerly living animals. His work along with Linnaeus's laid the foundations for modern biology.

Michael Denton discusses the contrast between what Darwin proposed and what these leaders of the natural sciences held to be true in his *Evolution: A Theory in Crisis*. In discussing the typological perception of nature, Denton shares why those leading naturalists thought differently. Denton notes, "...the founders of modern biology, those who discovered all the basic facts of comparative morphology upon which modern evolutionary biology is based, held nature to be fundamentally a discontinuum of isolated and unique types unbridged by transitional varieties..."[34] These men, who knew the world of living things and fossils as well as Darwin did, did not see evidence for the plastic evolving world that Darwin described. They did not see nature showing one species evolving from one to another. Linneaus's whole system is based on the opposite view. These founders of modern biology saw creatures "isolated" with no "transitional varieties."

For all these men the evidence showed that nature was not continuous but discontinuous. Nature's forms showed isolation, not connection, with no transitional forms. This verdict was the result of years of close study of geology, fossils, anatomy and living creatures. As Denton makes

34 Michael Denton, *Evolution: A Theory in Crisis*, Adler & Adler, Chevy Chase, MD, 1986, p. 100.

clear, what they saw was a typological view of nature in which, "there were absolute discontinuities between each class of organisms, that life was therefore fundamentally a discontinuous phenomenon and that sequential arrangements, whereby different classes were linked together or approached gradually through series of transitional forms, should be completely absent from the entire realm of nature. Typology contrasted completely with the idea of organic evolution."[35]

This is the view of nature that Darwin had to overcome when he finally published. And overcome it he did. But the story of the success of his ideas is not the story that most of us carry in our heads. Though his work met with acceptance by some, there were plenty who read his work carefully, admired his arguments and theory genuinely, but remained unconvinced.

Many of us believe that, with the publication of *Origin*, Darwin began a revolution in biology that swept the field that day and remained the accepted theory from his day to the present. Much like young boys who read the history of WWII and read the triumph of the Allies as inevitable, we may have in our minds that the triumph of Darwinian evolution was inevitable. But the triumph of evolutionary theory was not that simple or straightforward. For starters, his theory had to overcome the typological view of nature that Denton described. While his critics were duly impressed with Darwin's enormous knowledge and persuasive rhetoric, what was most needed to convince them was what was lacking: evidence.

"There can be no question that Darwin had nothing like sufficient evidence to establish his theory of evolution. Neither speciation nor even the most trivial type of evolution had ever actually been observed directly in nature. He provided no direct evidence that natural selection had ever caused any biological change in nature…the idea of evolution on a grand scale was entirely speculative, and Darwin was quite unable to demonstrate the 'infinite connecting links,' the existence of which he repeatedly admitted was crucial to his theory."[36] Denton notes also that Darwin was fully aware of these shortcomings. In the years following

---

35 Denton, *Evolution*, p. 98.

36 Denton, *Evolution*, p. 69.

the publication of his great book, he tried to meet them again and again in the subsequent editions.

Lack of evidence was not the only problem his critics made for him, but it was the greatest problem. The fossil record, his analogy from artificial selection to natural selection and conceptual difficulties were all part of the criticisms leveled against the theory. Thanks to David L. Hull's *Darwin and His Critics,* we have ready access to what Darwin's critics thought in the years immediately following the publication of *Origin.* But before we look at these reviews from Darwin's contemporaries, we need to deal with some preliminary matters.

Today's reader will benefit from an awareness of what many of Darwin's scientific colleagues thought about his theory and how they expressed their objections in the 1860s. It is particularly important to gain a sense of the nature of the objections Darwin encountered in the first years after the publications of *Origin.* One of the reasons the theory came under fire was because *Origin* was published at a time when the philosophy of science was in a state of transition. From Hull's viewpoint, the critics were sometimes wrong to evaluate the theory by the standards they used, and that Darwin's theory to be judged aright required a better understanding of the nature of Darwin's argument. The changes that occurred in science after Darwin's death made it easier to accept Darwin's theory.

Hull notes, "What was wrong with Darwin's theories? Was his methodology faulty? What is the nature of inductive formulation and proof which eluded Darwin? These are the questions that occur to the modern reader when first confronted by the reviews of the *Origin of Species,* and Darwin's later work especially the *Descent of Man.* …many of the reviews were by competent scientists honestly trying to evaluate a novel theory against the commonly accepted standards of scientific excellence, and evolutionary theory consistently came up wanting. …Darwin was caught in the middle of a great debate over some of the most fundamental issues in the philosophy of science…"[37]

Darwin was prepared for many to object to the content of his theory. He knew that Lamarck's theory of evolution with its forces driving spe-

37 David L. Hull, *Darwin and His Critics*, The University of Chicago Press, Chicago London, 1973, p. 14.

cies to perfection was held in contempt by most of his scientific peers. He had witnessed the scorn heaped on Robert Chamber's *Vestiges*. He had to make a case that would overcome the skepticism of his fellow naturalists. After years of work and writing with great skill and care, he believed his argument was presented in a way to satisfy his colleagues even if they did not accept his conclusions. For twenty years he assembled his evidence so that his argument would be well supported. He presented his case in such a way that he was sure it would pass muster from those whose opinion counted. He was therefore surprised and upset when so many of his colleagues objected, not just to his conclusions, but to his method.

Hull tells us that even those most likely to accept his work did not. "One would expect men like John Herschel, William Whewell, and John Stuart Mill, who had carefully investigated the nature of science as such to be better able to evaluate a new and novel theory than ordinary scientists and philosophers. Such was not the case. All three men rejected the tenets of evolutionary theory either entirely or in part. As the numerous comments scattered throughout the reviews of *Origin of Species* included in this anthology amply attest, one of the major objections to evolutionary theory was that it was not sufficiently 'inductive.' "[38] Objections like this from men of this stature posed a real difficulty. Darwin wrote to his friend Charles Lyell, "I have heard, by a roundabout channel, that Herschel says my book 'is the law of higgeldy-piggeldy'. What this exactly means I do not know, but it is evidently very contemptuous. If true, this is a great blow and discouragement." In the face of such a rejection by the most eminent philosopher-scientist of the century, it is easy to understand Darwin's pleasure when he discovered in an equally roundabout way that another great philosopher, John Stuart Mill, thought that his reasoning in the *Origin* was 'in the most exact accordance with the strict principles of logic.' On closer examination, however, Mill's endorsement can be seen to be not so reassuring. Darwin had properly used the Method of Hypothesis, but this method belonged to the logic of discovery, not proof. In spite of twenty years of labor, Darwin had failed to provide proof for his theory of evolution by natural selection."[39]

38 Hull, *Darwin*, p. viii.

39 Hull, *Darwin*, p. 7, 8.

Part of the evaluation of Darwin's novel ideas was the attempt by scientists to deal with the legacy of Francis Bacon. Bacon called for the work of scientists to be based on a proper inductive method of reasoning as opposed to a deductive method of reasoning. Bacon believed the deductive method of scientific reasoning, proceeding from general propositions to particulars, should be replaced by the inductive method, reasoning from particulars to generalizations. What concerns us now is not a thorough understanding of the importance of the difference between the two methods. Hull tells us Bacon's criticism of Aristotle's method of deduction is confused and that no one could properly distinguish and apply the two methods. What is important to understand is that, in Darwin's day, for one's theory to be seen as scientifically correct it had to be shown to be properly inductive. Darwin believed he had passed that test. The fact that his earlier theory on coral reefs, which was accepted and had not been faulted for improper inductive reasoning and was argued in the same manner as in *Origin,* shows the confusion at this time. It was very difficult for scientists to be consistent.

Hull maintains that, if evaluated on the basis of the hypothetico-deductive model, Darwin's theory passes muster. "Contemporary philosophers of science commonly concede that early philosophers of science were hopelessly confused on such key issues as the differences between discovery and proof, between deduction and induction, and between inference and concept formation."[40] Hull states that, among logical empiricists, the hypothetico-deductive (H-D) model is the most widely accepted. "…the chief virtue of the H-D model is that it recognizes the impossibility of absolute proof for any scientific theory. According to the method of 'inverse deduction' incorporated in the H-D model, a scientific theory can become increasingly more probable; that is all."[41]

Certainly if a theory need not be proved, at least in some sense, but just shown to be probable, then lowering the bar helps it meet the tests of science. Perhaps his contemporaries did ask for more than was reasonable in the way of proof, given the nature of what Darwin proposed. Darwin himself knew that his theory could not be "proved."

---

40 Hull, *Darwin*, p. 29.

41 Hull, *Darwin*, p. 30.

In a letter to H. G. Bronn, Darwin admitted, "You put very well and very fairly that I can in no one instance explain the course of modification in any particular sense. If evolution by natural selection was to be accepted at all, it would have to be chiefly from this view connecting under an intelligible point of view a host of facts. When we descend to details, we can prove that no one species has changed; nor can we prove that the supposed changes are beneficial, which is the groundwork of the theory. Nor can we explain why some species have changed and others have not."[42] If the devil is in the details, he is having a heyday.

One benefit a reader gains by reading Darwin's own words is to experience his refreshing honesty regarding the difficulties inherent in his theory—would that his modern-day defenders approached the issues with this same candor and goodwill.

Finding Darwin's evidence and argument convincing depends in part on your definition of science, your definition of the scientific method, as well as the evidence. Today's scientists are not of one mind on these matters. The reader today will hear statements on this subject from both ends of the spectrum. One man claims that the theory is now not just proved but established fact. Another will say that, not only is the theory not well established, but given the claims about events millions of years ago, it is silly to talk of proof and established fact.

Then there is the question of numbers. When Darwin first published, those who accepted his view were in the minority. Today, the numbers are reversed. Nearly all news you read regarding the controversy emphasizes that only a minority of thinkers are today skeptical. But one needs to read widely and to be as informed as possible to discover if the evidence and argument satisfy. It is difficult work, but patience and reading primary sources will help to develop one's own convictions. Beware of reading what others say about any given author, past or present. News reports seldom present the issues clearly or fairly. Given the herd mentality of most news reporting, the reader is doubly warned not to trust what you find in newspapers and news magazines.

Before we look at some reviews that Darwin had to contend with, let us end this introduction to them, with a final word from philosopher Charles Sanders Pierce. Hull reports that more than any other thinker,

42 Hull, *Darwin*, p. 32.

American philosopher, Charles Sanders Peirce saw that Darwin's argument was statistical in nature. Peirce was familiar with Darwin's type of argument from the work of Clausis and Maxwell. "The Darwinian controversy is, in large part, a question of logic. Mr. Darwin proposed to apply the statistical method to biology. The same thing has been done in a widely different branch of science, the theory of gasses… Darwin, while unable to say what the operation of a variation and natural selection in every individual case will be, demonstrates that in the long run they will adapt animals to their circumstances. Whether or not existing animal forms are due to such action, or what position the theory ought to take, forms the subject of a discussion in which questions of fact and questions of logic are curiously interlaced."[43] Peirce was a scientific thinker of the first rank, and one whom Hull states was nearly alone in 'approving the form of Darwin's theory.' His final thoughts: "What I mean is that his hypothesis, while without dispute one of the most ingenious and pretty ever devised, and while argued with a wealth of knowledge, a strength of logic, a charm of rhetoric, and above all with a certain magnetic genuineness that was almost irresistible, did not appear, at first, at all near to being proved, and to a sober mind its case looks less hopeful now than it did twenty years ago."[44]

In France, François Jules Picet noted that, while he could accept that natural selection resulted in limited change or variation in species, there was no reason to believe that this same action, no matter how much time was allowed, could create organs such as wings or eyes, nor was there evidence for this.

> Mr. Darwin accepts on the one hand the possibility of slight variations and on the other an immense series of centuries. He multiplies these two factors by each other and concludes that there are powerful and profound variations, not only in the external form but also in the most essential organs. Because these bold deductions do not appear to be justified by the facts, it is necessary to have a more powerful argument before accepting them. Nothing he has said proves to me that slight and superficial

43 Hull, *Darwin*, p. 33.

44 Hull, *Darwin*, p. 33, 34.

variations can in the long run change nature and degenerate in such grave modifications. I find nothing in the examples chosen by Mr. Darwin which authorizes me to believe that it is all a matter of degree, that just because one has demonstrated that in the course of some thousands of generations, the size, color, and the form of the beak can be modified, that the proportions of the limbs can be changed a little, etc., I am now forced to conclude that additional thousands of generations...will change gills into lungs, sprout wings, create an eye... All known facts demonstrate the contrary, that the prolonged influence of modifying causes has an action which is constantly restrained within by sufficiently confined limits. If we study the modifications brought about by domestication, remembering that these modifications are probably much more intense than those of natural selection, we cannot find a single example of an influence which has been able to modify the essential nature of an organ. Thus dogs, which have deviated more profoundly from their original state than any other domesticated animal and which are so diverse in their external form, retain an astonishing constancy of characters. If one may be permitted a slightly trivial observation, after all, dogs themselves are never confused. They always mount another dog, as different as the other dog may be from itself, whereas they would never mount a cat or a fox.

If the variations in all domesticated animals have been superficial and slight, then I must say that I know of nothing which proves nor an example which forces me to believe that the contrary took place in nature. Before I can accept Mr. Darwin's deductions, I must see for myself a known case of an important organ beginning to form or of a modification of some value in essential characters. As long as there is no proof that grave changes can be introduced regularly in order of direct generation, then I am drawn to the daily observation which leads me to believe the contrary.[45]

45 Hull, *Darwin*, p. 144–45.

Picet's critique points out that slight variations in external forms (think of the crops of Darwin's pigeons) does not mean one has demonstrated change in what he calls essential organs. Picet made a distinction between the change in the length of the beak of a bird and the sprouting of the wing of bird. One change is slight and superficial, and one is profound and essential. He noted that domestic or artificial selection is a more powerful force than natural selection. He further noted that domestic selection's power to change a species shows severe limits. One can breed for a faster horse from generation to generation, but there is a limit. We have yet to breed a horse that is faster than a cheetah. Finally, he touched on a problem that so many critics will note: Darwin's lack of examples. Darwin was well aware of this and freely acknowledged that he had no examples for his grand claim. In the parlance of the old TV commercial, Picet asks, "Where's the beef?"

William Hopkins was a mathematical coach whose students included Lord Kelvin. Hull noted that his critique was "especially damaging because it was impartial."

> …our author asserts that nature is always ready to render any variation permanent when it is for the good of the individual and its race. But what is *good* for an animal? Is it necessarily good that a sponge or a polyp should be promoted to the rank of an oyster, that the ambition of the oyster should aspire to the position of a cephalopod, or a cephalopod to that of a fish and so on? We should have thought it far more likely that the advantage which may thus be accorded to any organic being was intended for its good in the sphere in which it was born, but not to raise it above that sphere. At all events, we are sure that there can be no such probability on the side of our author's view as to constitute any positive *a priori* argument in its favour. He has a perfect right, however, to make such an hypothesis; but we maintain… that, admitting the existence of natural selection, there is no more reason *a priori* for believing that it has the power of raising an organic being from a lower to a higher stage of organic life, than there is for supposing its power restricted to those limits which the distinctive characters of natural species would impose upon

> it. The onus of proving the existence of such power must rest with him who asserts it.[46]

Hopkins notes that it makes perfect sense that natural selection should improve a species. Conceptually there is a great leap made from improving a species to changing a species into another. Of course natural selection could make an animal better from generation to generation. Improving a creature is one thing, but there is no reason to think improvement should promote an animal. A better sponge, yes, but a sponge to a polyp, no. Like Picet, Hopkins noted that there are limits to the power of artificial selection and believed natural selection would have less power, not more. Later in his critique, Hopkins, commenting on Darwin's propositions regarding hybridism "no mere turret, as it were, in our author's theoretical fabric—but the chief cornerstone which supports it," complains of the vagueness of Darwin's language.

> If we allow full weight to all our author's arguments in his chapter on Hybridism, we only arrive at the conclusion that natural selection *may* possibly produce changes of organization, which *may* have superintended the sterility of species; and that therefore the above proposition *may* be true, though not a single positive fact be adduced in proof of it. We confess that all the respect which we entertain for the author of these views has inspired us with no corresponding feeling towards this *may be* philosophy, which is content to substitute the merely possible for the probable, and which ignoring the responsibility of any approximation to rigorous demonstration in the establishment of its own theories, complacently assumes them to be right till they are rigorously proved to be wrong. When Newton in former times put forth his theory of gravitation, he did not call on philosophers to believe it, or else to show that it was wrong, but felt it incumbent on himself to prove that it was right. And this must be the rule for the philosopher who advances his theories on the phenomena of life, equally with him who speculates on the properties of inanimate matter. ...what we require is some proof of the fundamental propositions, founded on observation

46 Hull, *Darwin*, p. 251, 252.

or example or experiment. Till this or something equivalent to it be done, we maintain that no theory like Mr. Darwin's can have a real foundation to rest upon, or can ever secure the general convictions of philosophers in its favour.[47]

Hopkins here levels a blow to Darwin's whole way of persuading his reader. Darwin is so knowledgeable, is able to call up example after example, and writes so well that some readers may think he has proved a thing when all is assertions and assumptions. Hopkins calls Darwin's method of argument a "may be" philosophy. It is one thing to assert but entirely another to prove, or to cite the experiments or examples that demonstrate the truth of the claim. Remember, Hopkins is not asking for proof of variation in species or slight changes in externals, but for proof of Darwin's grand claim that these slight variations have the power to graduate one species to another. Later Hopkins notes, "our author makes at any time but little use of the verb, 'to prove' in any of its inflections. His formula is 'I am convinced,' I believe,' and not 'I have proved.' We are not finding fault with these more modest forms of expression; but we may be allowed, perhaps, to remark, that they are the formulae of a creed and not of a scientific theory. Views like this rest, in fact, on no demonstrative foundation, but as we conceive on *a priori* considerations…"[48]

Here again Hopkins has landed a telling blow to Darwin's method of persuasion. He states that he does not write like a scientist who lists scientific reasons for believing in his claim but writes like a shaman advocating his personal statement of belief. We will note here only one other objection from Hopkins about the lack of fossil evidence. "It is…in the records of geology that we must seek for the evidence of that continuity by which according to the theories we are considering, these changes must have been characterized. Now, all are agreed that paleontology affords no evidence of this continuity; and it is in the absence of this evidence that we meet with one of the gravest objections to any theory of continuous organic variation."[49]

Darwin freely acknowledged that what many saw as the gravest of problems for him was just that, a very grave problem for which he had

47 Hull, *Darwin*, p. 257, 258.

48 Hull, *Darwin*, p. 268.

49 Hull, *Darwin*, p. 258.

only the meekest of answers. We will have much more to say later about the subject of fossil evidence. For now, we will note that after Hopkins quotes Darwin's answer to this problem, he states: "It is by a vague hypothesis, entirely unsupported by facts, that our author meets a difficulty which appears to us, as it has appeared to many others, to be of the gravest magnitude.…Nor is this the only instance in which he seems to have adopted similar reasoning."[50]

Henry Charles Fleeming Jenkin, or Fleeming Jenkin, was a professor of engineering at University College London and the University of Edinburgh. He was a remarkably well-educated man, having studied in Frankfurt, Paris and Genoa. As an engineer he worked with Lord Kelvin in laying the first transatlantic cable. In his criticism of Darwin's theory, he echoed Hopkins complaint of Darwin's "may be philosophy." Early in his review Jenkin noted that experience with domestic variation teaches that variations to a species have limits and over time the degree of variation diminishes. This was common knowledge of domestic breeders. He likened Darwin's theory of how variations might work to that of a cannonball fired from a cannon. "To conclude that because we observe that a cannonball has traversed a mile in a minute, therefore in an hour it will be sixty miles off, and in the course of ages that it will reach the fixed stars. This really might be the conclusion drawn by a savage…"[51]

He noted that when breeding for a certain trait, speed in horses for instance, that the earlier generations show a more marked improvement than later ones. From generation to generation, the degree of improvement decreases. Breeders come to a point where they reach a natural limit and how far they can push a certain trait. "As matter of experience…have our racers improved in speed by one part in a thousand during the last twenty generations? Could we double the speed of a carthorse in twenty generations? Here is the analogy with our cannonball: the rate of variation in a given direction is not constant, is not erratic; it is a constantly diminishing rate, tending therefore to a limit."[52]

Jenkin also finds fault with Darwin's explanation of how natural selection would overcome the tendency of offspring to revert to type, to

50 Hull, *Darwin*, p. 263, 264.

51 Hull, *Darwin*, p. 306.

52 Hull, *Darwin*, p. 307.

be just like their parents. With domestic selection, breeders are able to control the breeding parents to insure that the trait the breeders want has the best chance of continuing in the offspring. If they did not, the trait, such as speed, would be lost. If a colt has inherited fast legs, the breeder does not want that colt to breed with an inferior mate. The breeder will control the breeding to increase the chance of passing the trait to the next generation. From generation to generation he must keep the trait he is breeding for "on line." He would want to match the fastest colt to the fastest philly. If offspring are free to mate with any animal generation from generation, then the trait desired will be lost. The controls of the breeder are crucial to preventing the next generation from reverting to type. Without control from breeders the subsequent generations produce average offspring. So not only must natural selection maintain its strength, it must also keep the trait that will change the animal "on line" from generation to generation to generation. Man can guide or override what occurs in nature. But without his intelligent interference, animals revert to being like their parents. Where was the evidence that nature working without imposed restraints could prevent an animal from reverting to type?

> We hope this argument is now plain…if in every case we find that deviation from an average individual can be rapidly effected at first, and that the rate of deviation steadily diminishes till it reaches an almost imperceptible amount, then we are as much entitled to assume a limit to the possible deviation as we are to the progress of a cannonball from a knowledge of the law of diminution in its speed. The limit to the variation of species seems to be established for all cases of man's selection. What argument does Darwin offer showing that the law of variation will be different when the variation occurs slowly, not rapidly? The law may be different, but is there any experimental ground for believing that it *is* different? Darwin says (p. 153), 'The struggle between natural selection, on the one hand, and the tendency to reversion and variability on the other hand, will in the course of time cease, and that the most abnormally developed organs may be made constant, I can see no reason to doubt.' But what reason have we to believe this? Darwin says

> the variability will disappear by the continued rejection by the individuals tending to revert to a former condition; but is there any experimental ground for believing that the variability *will* disappear; and, secondly, if the variety can become fixed, that it will in time become ready to vary still more in the original direction, passing that limit which we think has just been shown to exist in the case of man's selection?[53]

Again, a weakness of Darwin's foundational analogy is exposed. For men like Hopkins and Jenkin, the analogy creates more problems than it solves. But this analogy was central to Darwin's theory of natural selection. Alfred Russell Wallace had come to much the same theory of how species came to be. His manuscript had forced Darwin to go public. He believed that Darwin was foolish to argue for origins in nature based on domestic selection. Wallace believed that the analogy Darwin thought crucial to his theory was in fact detrimental to the theory.

Finally, from Jenkin, his problem with Darwin's lack of hard science to support the theory:

> The chief argument used to establish the theory rests on conjecture. Beasts may have varied; variations may have accumulated; they may have become permanent; continents may have arisen or sunk, and seas and winds been so arranged as to dispose of animals just as we find them, now spreading widely, now confining it to one Galapagos Island. There may be records of infinitely more animals than we know of in geological formations yet unexplored. Myriads of species differing little from those we know to have been preserved, may actually not have been preserved at all. There may have been an inhabited world for ages before the earliest known geological strata. The world may indeed have been inhabited for an indefinite time; even the geological observations may perhaps give a most insufficient idea of the enormous times which separated one formation from another; the peculiarities of hybrids may result from accidental differences between the parents, not from what have been called specific differences.

53 Hull, *Darwin*, p. 308, 309.

> We are asked to believe all these maybe's happening on an enormous scale, in order that we may believe the final Darwinian 'maybe,' as to the origin of species. The general form of his argument is as follows:—All these things may have been, therefore my theory is possible, and since my theory is a possible, all those hypotheses which it requires are rendered probable. There is little direct evidence that any of these maybe's actually *have been.*[54]

Thomas Vernon Wollaston (1821–1878), a prominent English entomologist, concluded his review of the theory by saying, "Although we have felt compelled to say thus much against the theory so ably pleaded for in Mr. Darwin's book, we repeat that, *in a very limited sense indeed,* there seems no reason why the theory may not be a sound one; but at present, even to that extent, it remains to be substantiated."[55] A few pages back, Wollaston noted not just the lack of evidence to substantiate Darwin's claims, but that the evidence was all for the opposite of his claims.

> He states his difficulties with honesty, precision, and clearness, and sometimes (as it appears to us) even exposes them more than is necessary—to his own disadvantage… in spite of this candour on his part (a candour which is so manly and outspoken as almost to 'cover a multitude of sins'), we thought *any* of them satisfactorily replied to. There is a clever and ingenious pleading for them all; but, if we look back into the volume, we find (to use the mildest expression) that, each, in its turn, has been left in doubt—awaiting further evidence. So that, until this is forthcoming, we cannot but feel that, whilst the theories are in one direction (and made to dovetail into each other) the great body of facts is unquestionably on the opposite side. More especially will this apply to that gravest of all objections (as Mr. Darwin frankly admits), *the thorough and complete absence…* of that countless host of transitional links which, on the 'natural selection' theory, must certainly have existed at one period

54 Hull, *Darwin*, p. 338, 339.

55 Hull, *Darwin*, p. 139.

or another of the world's history. ...at present it is absolutely certain that we have not so much as a shadow of evidence either that they do exist, or have ever existed. ...To an uneducated eye, which views things only in the mass, this may not appear *prima facie* evident; but those who have worked closest and longest at details, in the open face of nature, know that it is true. ...the plain fact remains, that, on a broad scale, more or less abrupt and well-defined forms alone have as yet been discovered, and that they do *not* shade off into each other by that legion of osculant infinitesimal links on which the very life, as it were, of the ingenious theory mainly depends.[56]

56 Hull, *Darwin*, p. 136.

CHAPTER 8

THE TRIUMPH OF EVOLUTION

...after the Oxford debate...(Huxley) opted for apes. Apes and evolution let him channel his gifts into a single, high-blasting campaign, one that simultaneously allowed him to further his work in the biological sciences, criticize theological authority, advance the cause of young professionals like himself, tackle enemies like Owen on their home territory, promote a naturalistic approach to the living kingdom, and feed his lust for life and combat.

—From *Charles Darwin: The Power of Place*, by Janet Browne.

SOME READERS MAY VIEW THESE objections as substantial; others may not. It may be difficult for some to see them as valid having grown up in a world in which Darwinism permeates all categories of thought. Some readers may read the preceding and think: so what? We know who won this debate. Darwin won hands down. The debate was settled years ago and has since been confirmed by science time and time again.

But to view these criticisms in this way is to misunderstand in great part the story of Darwin's triumph. Given the weaknesses of *Origin* how did Darwin's ideas instigate a revolution in Western thought in such a short time? Before the end of the nineteenth century, Darwin's theory was on track to dominate scientific thought and do so with the blessing of the Anglican Church and other liberal church leaders. Darwin's body would be lying in Westminster Abbey. The man who did more than Freud or Marx to undermine the Christian view of the world was honored in burial in the Abbey! Given the revolutionary nature of his ideas, this triumph in such a short time is an extraordinary event. What were some of the key factors that gave Darwin the victory?

First, Darwin's ideas succeeded because the time was right. Some sort of evolutionary theory had been buzzing around since the first half of the nineteenth century. Darwin's mentor as a teenager in Edinburgh, Edmond Grant, was convinced that all of nature could be accounted for by strictly natural processes. In Darwin's youth, this was radical thinking, but over the years ideas about evolution had become less so.

For many, evolution was very appealing. Progressive thinkers in Darwin's day were convinced trends and developments in nature and society were all upward. Victorians seemed to be surrounded by progress. The growth of Great Britain's industrial might and global reach confirmed this. The economic and technological progress was so striking it was celebrated at the Crystal Palace. Here, one could see it all at display after display. The Industrial Revolution was not limited to Great Britain, but an Englishman might be forgiven for thinking that his countrymen were at the top of the scale. However, it is important to point out that Darwin's bleak vision did *not* endorse this view of progress at all. His was a thoroughly Malthusian dog-eat-dog vision. There was no such thing as progress; indeed, in Darwin's vision, the term progress was meaningless. For Darwin, there were winners and losers. The strong won and the weak lost. But so eager were so many to embrace a totally naturalistic explanation of all things that they easily overlooked this aspect of his theory and got on with the work of installing evolution as a new paradigm for scientific thought.

There was the work by authors that preceded Darwin, most notably Lamarck and Chambers. Jean Baptiste Lamarck's version of evolution postulated an inner force that guided living things to greater and greater development. There were many evolutionists before Darwin of the Lamarckian kind, and for them Darwin confirmed and provided a greater scientific basis for their belief. Darwin himself could never sign on to Lamarck's idea because his purpose was to explain it all without resorting to any unaccounted-for "inner force." Nevertheless, Lamarck did much to preparc others to accept Darwinian evolution when that was presented to them. Robert Chamber's *Vestiges,* the wild panoramic tale that began with the birth of stars and ended with humankind, was a sensation in its day. Though his work had little science in it, his story was enough for many to cross the bridge from the traditional view of creation to a naturalistic one.

Not only were the upper classes primed for an evolutionary view of the world, the lower classes were as well. With the upheaval caused by the Industrial Revolution, the lower classes in Great Britain were becoming less accepting of their place in society. They were eager to believe in the ideas of progress and upward mobility. A picture of nature that showed the lower evolving into the higher meant they were part of nature's great progressive ascent and would no longer necessarily be confined to their low station. Defending the traditional Biblical view of the world were those at the opposite end of British society: the graduates of Oxford and Cambridge who ruled England. In a debate over the merits of Darwin's theory it is easy to understand that the members of the lower class would be eager to hear that the traditional creation story which—according to their betters, assigned them to their place—was in error, and that they were *not* ordained by God to forever be miners and maids. In this revolutionary theory of life, they were given to understand that biology taught that the lower evolved into the higher. The theory held that man and nature are one. They could therefore expect to rise in society because that was how nature worked.

Then there was Darwin's great book. If *Origin* had its faults, it surely had its strengths. In it he proposed that his natural explanation of origins better accounted for rudimentary organs, the progression of the fossil record, and the geographic distribution of species than did the traditional story of creation. Plus, he provided the all-important completely natural mechanism for it all: natural selection. No need now for some unexplainable inner force like Lamarck's. Now one had the whole package—all-natural, neat and clean. Nature had been "brought to heel." He presented simple concepts that followed logically: there is variation in species, positive variations can be inherited, the inherited variations could confer an advantage in the struggle for life thus improving the species. Given enough time natural selection could not only improve the species but create new organs and body forms and even new species. It was, as he said, one long-sustained argument. The evidence was all circumstantial evidence, but it was a lot of circumstantial evidence. He brought fact upon fact, upon example, page after page after page, until the reader was worn down. Even for those who were not convinced that natural selection could do all that he claimed it could, the opportunity presented to believe in an all-natural explanation for the origin of species was too great to pass up. Now all was explained, and all explained by science, with no mystery, and no need for any creator or designer. For many, he made an offer they couldn't refuse.

Origin also benefited from the effect of years of German Biblical criticism. The fruition of years of Enlightenment thinking, the divines from Germany introduced into Protestant theology skepticism that undermined the trustworthiness of the Bible. Anything involving miracles and the supernatural became suspect. Old Testament stories like Noah and the great flood, the parting of the Red Sea, and New Testament stories like the feeding of the five thousand, the virgin birth, and of course the resurrection of Jesus, came to be seen as unhistorical or unscientific. The Bible presented God as the creator and sustainer of the natural world. Having the authority of the Bible undermined helped pave the way for an alternative creation story, one in which God had no role and a self-sustaining nature did it all.

Then there was Darwin's own personal group of friends. He had carefully crafted a network over the years now in place and ready to do what he needed done to insure the success of this theory. Of this group Janet Browne says the following:

> …It was to Darwin's friends that the first wave of positive responses must be attributed. For it was obvious that Darwin's theories were as useful to them as they were to his theories. Over the following decades, Darwin's defenders came to occupy influential niches in British and American intellectual life. Together, these men would also control the scientific media of the day, especially the important journals, and channel their other writings through carefully chosen publishers—Murray, Macmillan, Youmans and Appleton. Towards the end they were everywhere, in the House of Parliament, the Anglican

> Church, the universities, government offices, colonial service, the aristocracy, the navy, the law and medical practice; in Britain and overseas. As a group that worked as a group, they were impressive. Their ascendancy proved decisive, both for themselves and for Darwin.[57]

Part of this network was the so-called X Club that included Joseph Hooker, John Lubbock, Herbert Spencer, and John Tyndall.

Joseph Hooker (1830–1890) was the man with whom Darwin first shared his transmutational heresy. A botanist and the director of Kew Gardens, he had slowly over the years become a believer in Darwin's theory, as well as a close personal friend. Herbert Spencer (1820–1903) was the radical social philosopher who, before Darwin published, applied a Lamarckian view of biological progress to social progress. After Darwin published, Spencer became a devoted Darwinian. It was Spencer who coined the phrase "the survival of the fittest." Irishman John Tyndall (1820–1893), geologist, mathematician, and lecturer, declared at an X Club gathering, "We shall wrest from theology the entire domain of cosmological theory."

Charles Lyell (1797–1875) impacted a generation of scientists' and thinkers' writing on geology. He traveled in the very upper tier of British society. Asa Gray of Harvard became the conduit of Darwinian theory to North America. Though he was a reluctant convert and would never be a full disciple of Darwin, he would promote and protect the theory on the western side of the Atlantic from his powerful post at Harvard.

Finally, there was Thomas Huxley (1825–1895), the man in Great Britain who would do more than any other individual to use Darwin's theory to bring about a revolution. His influence is hard to exaggerate. Huxley not only did more than any other spokesman to establish the theory; more importantly, he would be the man who would do so much to establish the tenor of the debate. Unlike the man whose theory he would champion—who was born into wealth and privilege—Huxley was born into poverty. He reached the positions he held in British science by pulling himself up by his bootstraps. He carried a deep resentment toward the Oxbridge graduates who dominated British life and was

57 Janet Browne, *Charles Darwin: The Power of Place*, Alfred A. Knopf, New York, 2002, p. 129.

determined to wrest from them the control they exercised on science. Before *Origin,* he was already committed to a naturalistic worldview, but now Darwin with his theory of natural selection had placed in his hands a tool he could use to smash the establishment. And smash them he would, even though at the time he began his crusade he did not believe that Darwin got it exactly right with natural selection. In fact, he had written to Darwin warning him that he was creating for himself a great difficulty maintaining that evolution occurred by only slight modifications. At the time Huxley began his crusade, his own conception of the origin of species involved certain unalterable types. Darwin's slow progression of slight modifications from just one or a few ancestors was so far removed from his own view that he thought natural selection an error and one Darwin would never be able to overcome.

Surely this was a curious state of affairs. Natural selection was Darwin's great gift to those longing to believe in some sort of viable naturalistic evolution. Natural selection was the child of his pride. It made it possible for a totally natural explanation of the origin of all species. So how could the man, known later as "Darwin's Bulldog," not believe in the cornerstone of the theory and yet be its greatest champion? He could do so because for Huxley, as for many others as well, the mechanism was not important. Answering the likes of Picet or Jenkin was not important. What Darwin was most proud of, the very science that made his theory work, was for Huxley easily expendable. The truth was, for people like Huxley and for many others anxious to throw out the traditional view of creation, Darwin had shown that evolution was true. If he was wrong about the how, that was not a problem. The pages of examples and facts Darwin had gathered from all his years of thought and study had done their work. Though there was no real evidence for Darwin's claims, his book convinced many by the wealth of material. A perceptive reader like Thomas Huxley could see that Darwin's mechanism was faulty (Huxley did not believe that slight modifications over time could produce what Darwin claimed they produced), and yet he is sure that Darwin had shown evolution to be true.

The strategy for Huxley was to get on with winning the argument and to wrest power from the establishment. Paint those who defended the traditional view as opponents of progress and defenders of privilege.

Use ridicule and sarcasm and demonize those who hold the opposite view and a real debate on the merits of the theory will not take place. It will not need to take place. Claim now that the thing was proved. Claim that those who did not believe were old fogies and obstructionists using religion to stop the advance of modern science. That was the trick. Huxley and the X Club would assume the role of Galileo and anyone who disagreed was like the church of old. The opposition could be painted as dishonest men of bad faith and false motives. With the debate framed in this way, and with a man of Huxley's skill and ferocity ready to use any rhetorical weapon to smite the opposition hip and thigh, it is no surprise that Darwin's side won.

What Huxley won was not just the vindication of Darwin's ideas, but the creation of a new class of scientists. Not only would Darwin's theory reign triumphant but he would make it possible for men like himself, not born of wealth and educated at Oxbridge, to be leaders in British science. In fact, it was through the efforts of Huxley and his contemporaries that the idea of science as a profession was born. Before this victory, science had been for the most part the privilege of men like Charles Darwin, men born to wealth and high social position. With the revolution of Darwin's idea came a revolution of what it took to be a scientist.

By the time of Darwin's death in 1882 his ideas were changing not only biology but were impacting all disciplines of science and philosophy. He had given the world an alternative creation story. A new scientific story. In this version, life begins by accident in some pond of primordial soup, and through struggle and chance over millennia one or a few ancestral forms produce all the life-forms we have today. As life-forms struggle to find their niche in order to survive they are forced to change or become extinct. This struggle produces new species. There is no purpose or reason for anything. All is the result of chance and you, as a human being, are for the moment, at the top of nature's pyramid.

As more scientists and thinkers signed on to Darwinism, those who did not were made to look more and more out of it. The business of discrediting Holy Scriptures continued a pace. To believe that God, or any agency outside of nature, had something to do with the origin of species was to be associated with a book that was said to be a collection of myths and fables. Such a book could have nothing to say about how life began or developed. The Jewish carpenter from Nazareth was seen more and more as a figure from a fable. Jesus had taught nice and useful things, but he was no longer a figure of history like his contemporary Caesar Augustus.

With Darwinians occupying the progressive and scientific ground, one was automatically set for ridicule if not on board. Thanks in part to the tenor of the debate, as things developed in the late nineteenth and early twentieth centuries, a phenomenon like the one described by the great Danish storyteller Hans Christian Anderson in "The Emperor's New Clothes," fell upon the disputants.

"The Emperor's New Clothes" tells the story of a monarch who cared only for the newest and finest clothes—in short, he is a fashion nut. Into his town come two swindlers who claim they can "weave the most beautiful stuff imaginable," but that clothes made of this stuff had the peculiar quality of becoming invisible to every person who was not fit for the office he held, or if impossibly dull.

At the time *Origin* was published, there were many intellectuals who wanted more than anything to have established a thoroughly naturalistic theory of life. For many this desire was so great that it did not matter if they did not properly understand Darwin, or even if they misunderstood Darwin. Here was the theory they had been looking for. With it, they could overthrow the old paradigm and cast out the traditional religionists while creating a new class of arbiters of what was true.

In Anderson's story, while the weaving swindlers are at work, the emperor sends a trusted minister to check on the clothes. He does not go himself for fear that when he gets to the loom he will not see anything, and he will be shown unfit for his office or too dull to see. When the minister arrives he of course sees nothing but cannot admit it. A second court official is sent to look in on things with the same result. Finally, the emperor screws up his courage in order to see for himself the work in progress. Yikes! His worst fears are confirmed. He does not see anything on the loom. What to do?—fake it. Now that the program is in place, one cannot afford to be one who does not see.

This was the kind of pressure facing any who would not assent to evolution. In America, Asa Gray's counterpart at Yale, Louis Agassiz did not accept Darwin's theory, and the once leading American naturalist became an increasingly marginalized figure in American science. Not to be on board with the new paradigm showed how unfit you were for your position. As the Darwinian tide swept on, if a man voiced his lack of faith in Darwin's theory it was tantamount to stating he did not "see" and therefore was unfit for science. Anderson's story ends with the emperor trying out the new clothes in a procession where a child states the obvious, "he's not wearing anything!" Once the child has voiced the truth, the townspeople, who had been going along with the sham, are free to acknowledge what their eyes have shown them all along to be the truth. But there was no such clarifying voice for people like Louis Agassiz.

Darwin's great book was one long speculative argument. He knew his case was far from demonstrated. The case was circumstantial and he was willing to acknowledge he asked a lot of his readers. He famously said that when he thought about how natural selection might produce a marvel like the eye, it made him shudder. He understood there were great difficulties in accepting his work.

But in the swirl of public events, books, debates, pamphlets and reviews, his theory took on a character like the clothes of the emperor. Even though it was not in any way proved, nor could it be proved, it became a thing proved in the public's mind. Those who did not see this paid the price of being cast out of the new scientific community. If one did not "see" evolution, one was no longer a scientist. The subtitle of Desmond's biography of Thomas Huxley is *"From Darwin's Bulldog to Evolution's High Priest."* That was the kind of power the believers of Darwin had, the power of a priest. They came to wield that power like a priest looking for heresy. To this day the defenders of Darwin's view of the world often defend it more like religionists than scientists.

By the turn of the century, though the problem of inherited traits was taking a lot of shine off the theory, it still had great momentum among many thinkers. For some, the thing was as good as proven; but what thing exactly? How had Darwin's speculations about origins and his extrapolation from domestic selection to natural selection been shown to be true? In fact, it had not been. Following the publication of *Origin,* examples of natural selection's power were found in the mimicry of butterflies and fossil bones. Examples like this were said to have confirmed the theory and that was deemed proof enough. But mimicry in butterflies, however it came to be, is not on the scale of change of reptiles into birds, which was the grand claim of the theory. This kind of claim remained undemonstrated. It was a claim that could not be demonstrated. The only key from the past, the fossil record as it was known, contradicted the theory. The important point is that none of this mattered. The theory was thought to be proven, claimed to be proven by so many with such authority, that for all intents and purposes it was proven.

So here we have it: a theory that convinced people of the fact of evolution, but its mechanism, natural selection, was not fully accepted. The problem of natural selection accounting for the origin of species was so great that Darwin had to retreat from it in one edition after another of *Origin*. Unlike his "bulldog" advocate Thomas Huxley, Darwin was an honest and considerate gentleman. With each new edition of his book, he tried to answer his critics regarding the power of natural selection to do what he claimed for it. By the sixth edition, he had dodged and feinted so much that he had greatly diminished the role he had at first assigned to it. The retreating was making a hash of his theory. Lauren Eiseley notes in *Darwin's Century*:

> A close examination of the last edition of the *Origin* reveals that in attempting on scattered pages to meet the objections launched against his theory the much-laboured-upon volume had become contradictory… The last repairs to the *Origin* reveal, both in connection with Lord Kelvin and Jenkin, how very shaky Darwin's theoretical structure had become. His gracious ability to compromise had produced some striking inconsistencies. His book was already a classic, however, and these deviations for the most part passed unnoticed even by his enemies.[58]

By the turn of the century scientists were falling back on some combination of Darwin's and Lamarck's theories. Lamarck postulated an agency, a life force of some kind, that drove creatures to develop and the changes were inherited by succeeding generations. Darwin's natural selection eliminated the need for this unnamed and unscientific force, but now with natural selection insecure Lamarckianism was making a comeback.

---

58 Loren Eiseley, *Darwin's Century*, Anchor Books, New York, 1961, p. 242.

Lamarck's theory had been around long before Darwin published. In *Origin,* Darwin brought to naturalists a more scientific theory. Why then were scientists now returning to Lamarck, and why did they finally settle on Darwin? Part of the reason for this was their faulty understanding of inherited characteristics.

We have encountered this problem before in the critics' concern that the theory could not explain why, in the wild, adaptations would not be lost as animals reverted to type. In Darwin's scheme, slight, beneficial variations are continuously passed on from one generation to another. But based on how they understood inheritance to work, this would fail due to the swamping effect. Let us imagine a rabbit that is faster than his fellows and that speed is a beneficial trait allowing him to better escape predators. Let us assume he can pass this trait on to offspring, but the faster offspring can breed with any part of the rabbit population. For the trait to continue the fast rabbit needs to mate with another speedy rabbit, but chances are he might just as well hook up with a real slowpoke. When this happens the trait is diluted and through the succeeding generations the trait gets swamped. In Darwin's time, naturalists thought inherited traits were blended. The offspring inherit speed from one parent but slowness from the other. With these two traits blended, the speedy grandpa rabbit two generations back does not see his speed passed on to following generations.

Lamarck, on the other hand, believed that an acquired trait would simply be passed on plain and simple. The speedy rabbit passed on his speed to his children and they did the same. Lamarck's famous example was his explanation for the long neck of the giraffe. In Lamarck's version the giraffe's neck had been of normal length at one time. Over time as giraffes stretched their necks to reach the leaves on the upper branches of trees, the giraffe's neck got longer. The next generation stretched the neck more, and so on and so on, until today we see the famous long neck of the giraffe. Lamarck believed that some vital force accounted for the changes in species.

The great problem with Lamarck's theory was there was no evidence that this was so. Experiments by proponents and opponents failed to demonstrate his proposition. But doubts about natural selection had even Darwin becoming a little Lamarckian in his attempts to fend off his critics.

The rescue for Darwin's ideas came from the Dutch botanist Hugo DeVries who postulated a mutation theory, and the monk Gregor Mendel, who discovered how traits were truly passed on. Their work showed the old idea of blending to be an error. DeVries dropped any Lamarckian idea of acquired traits being passed on; only an inborn trait could do so. He proposed that a dramatic change or mutation could occur, not in just an individual creature, but in a small breeding group where the trait would not be swamped. Natural selection would preserve the innovation. With the discovery of Mendel's work scientists knew that the idea of blending was wrong. A trait could be recessive in one generation, then appear in the next. Mendel's accurate description of how traits were inherited and DeVries's mutation idea had solved the swamping dilemma. In the following years, R. A. Fisher, J.B.S. Haldane, Theodosius Dobzhansky, Sewall Wright, Julian Huxley, Ernst Mayr, and George Gaylord Simpson, using Mendel's genetic work, would develop what is known as the Neo-Darwinian synthesis. This modified Darwinian theory is today the basic framework for how natural selection is said to work. But before we examine the status of the theory at the beginning of the twenty-first century, let us take a look at an event that occurred in America and the influence its popular interpretation has had on the debate over evolution.

CHAPTER 9

THE TRIUMPH OF PROPOGANDA

It is important to the concept of the play that
the town is visible always, looming there,
as much on trial as the individual defendant.

—Stage directions for Act One Scene 1, *Inherit the Wind,* by Jerome Lawrence and Robert E. Lee.

MOVIES HAVE A DEEP IMPACT on people's understanding of history. Many people adopt their position on controversial issues by what they have learned from movies and popular culture. Take Oliver Stone's film *JFK.* I have seen the film three times. It is stunning storytelling, and while watching it I am caught up in Stone's world of conspiracy. But I have also read books about the assassination of President Kennedy and I know that most of Stone's film is simply malarkey. If right now we could arrange for an accurate count, I would wager that the number of people who have seen *JFK* far outweighs the number that have read Gerald Posner's *Case Closed.* Author Vincent Bugliosi's *Reclaiming History: The Assassination of President John F. Kennedy* has added greater testimony to the truth that Lee Harvey Oswald acted alone. But for many, it is not the work of the careful researcher who has the most impact on a subject but rather the inflammatory propagandist. Mr. Bugliosi's book is an exhaustive treatment of the assassination, and for that reason alone it will not be read by many who this coming weekend will watch *JFK* on TV and learn history from Hollywood.

In 1925 there was a momentous trial in the small town of Dayton, Tennessee. Here are some of the facts we know about that trial. Angry Christian citizens removed a biology teacher from the classroom and got him thrown in jail when they learned he was teaching their children Darwin's theory of evolution. These Christians are in lynch-mob mode and are the most ignorant, mean, bigoted lot you can imagine. The national figure they welcome to their community to help prosecute the hapless schoolteacher is a man whose low character matches their own. He is a pompous windbag, nasty, stupid, duplicitous and lest any character flaw be lacking, a glutton. The Christians in the town are so angry and violent that the teacher's life is in jeopardy while awaiting trial. But the problem with these facts is that they are not facts at all. None of the preceding is true. But this is what many believe to be true because they have learned about the Scopes trial, not from history but from the play *Inherit the Wind,* or from of the film versions of the play. The most famous version stars Spencer Tracy and Frederic March.

The play's authors, Jerome Lawrence and Robert E. Lee, were moved to conceive their play in response to the dangers they saw in the work of Senator Joseph McCarthy. Their immediate concern had nothing to do with Darwin, but with the threat to individual liberty they perceived from the Senator's tactics. They used the events of Dayton, Tennessee in 1925 to tell their own story. They were not retelling the story of the actual trial. The villain in their play, Matthew Harrison Brady/William Jennings Bryan, lacks any redeeming quality. All he is missing to define him as the bad guy is a black hat. Their hero, by contrast, is the most gentle and reasonable of men who, even if one disagrees with his views on religion, one is compelled to like. For the playwrights there are no shades of gray; one man wears a white hat, the other a black one. There are no nuances about the issues at stake either. Their story is a clash between ignorant fearful Christians of fictional Hillsboro and progressive enlightened champions of free speech. This may or may not make for good theater, but it certainly does great injustice to the real men and women caught up in the trial in 1925.

To learn what really happened at the Scopes trial one should turn to Edward Larson's Pulitzer Prize winning book *Summer for the Gods*. For starters we learn how there came to be such a trial in a small mountain town situated between Knoxville and Chattanooga. In the 1920s the ACLU was busy protecting the individual's right of free speech from government policies and majority rule. When the state of Tennessee passed a law prohibiting the teaching of evolution in the public schools, they saw a threat to individual free-speech rights. They wanted a test case and ran ads offering their aid to anyone willing to challenge the law. A transplanted New Yorker living in Dayton, George Rappleyea, was looking for a way to get little Dayton, Tennessee, publicity. He disliked the law, read the ACLU's ad, and according to the best sources headed for Fred Robinson's drugstore. Robinson's store, equipped with a soda fountain, was a local meeting place; he also chaired the Rhea County school board. Robinson and Walter White, a former Republican state senator, listened to Rappleyea's idea for getting their little town on the map. Both were soon on board. Rappleyea let them know he had connections to New York and the ACLU. Add to this drugstore gang Dayton's city attorneys, Herbert E. Hicks and Mr. Sue K. Hicks. They said they were willing to prosecute. What was needed now was a teacher to prosecute. They next solicited John T. Scopes, a general science teacher and part-time football coach, who had filled in for the regular biology teacher at the end of the school year. Having filled in for the regular biology teacher, he could admit to breaking the law. The group asked Scopes to be part of a test case. Scopes said he was willing. Mr. Hicks said he was willing to prosecute. Rappleyea called a justice of the peace who swore out a warrant. Scopes then left to play tennis, and Rappleyea called the ACLU. So much for the angry mob removing a teacher from his classroom! The group that was eager to get publicity for Dayton was about to succeed beyond their wildest dreams.

Such are the origins of the infamous Scopes monkey trial. As for the town of Dayton and how it is represented in the play, it is worth quoting from Larson:

> Journalists covering the trial commented on Dayton's distinctiveness from the outset. None found rancor. Even the hypercynical Mencken wrote in his first report from Dayton, "The

> town, I confess, greatly surprised me. I expected to find a squalid Southern village, with darkies snoozing on houseblocks, pigs routing under houses and the inhabitants full of hookworm and malaria. What I found was a country town full of charm and even beauty… Nor is there any evidence in the town of that poisonous spirit which usually shows itself where Christian men gather to defend the great doctrines of their faith," Mencken added. "The basic issues of the case, indeed, seems to be very little discussed at Dayton. What interests everyone is its strategy."[59] Within a few days, Mencken realized that their primary interest was in good publicity and quietly skipped out of town before his increasingly caustic reports got him in too much trouble.[60]

When a man like H. L. Mencken finds your town full of charm and beauty and no one finds the rancor they expect given what is at stake, then the citizens of your town must be people of exceptionally fine and noble character and nothing like the citizens of fictional Hillsboro. However, in the play when Henry Drummond makes his entrance, E. K. Hornbeck (H. L. Menken) greets him with the words, "Welcome to hell." For Lawrence and Lee, the place is so foul only hell will do.

At the moment in the play when they learn that Henry Drummond/ Clarence Darrow will be coming to help the defense, the preacher Jeremiah Brown, the local pastor, calls Drummond a "vicious, godless man," an "agent of darkness," and declares that they will not let him enter the town. Here again from *The Summer for The Gods*: "Despite… efforts to sensationalize the contrast between Darrow and Daytonians, townspeople immediately took to the great agnostic. 'He arrived wearing a straw katy, his coat open in a gesture of summer casualness,' Scopes later recalled. 'It was easy to like him. He drawled comfortably and hadn't any airs. He gave the impression he might have grown up in Dayton, just an unpolished, casual country lawyer, so ordinary did he act.' Darrow sized up the town, conferred with Scopes, and met the press. The progressive Dayton Club hosted a banquet in his honor, which gave him a formal opportunity to explain his views on evolution and religion.

59 Original footnote from *Summer for the Gods*: Mencken, "Monkey Trials," 36-37 (Reprint of 9 July 1925 article).

60 Edward Larson, *Summer for the Gods*, Basic Books, New York, 1997, p. 93

'People of Dayton like his personality and think he is a great man,' Sue Hicks reported to Bryan in a letter, 'but they are all shaking their heads about his beliefs.'"[61]

And so it goes. Like all those who "know" that President Kennedy died because of a conspiracy, most people believe the citizens of Dayton would have lynched Scopes if they had had the chance. The truth could hardly be more contrary. Scopes never spent time in a jail cell, but it is the play's version of events that carries the day. We have already seen how the authors portray the stand-in for William Jennings Bryan, but a word needs to be said about his motives. Bryan's political philosophy was based on his trust in the people of the country and their wisdom. Of course, he opposed the teaching of Darwin's theory on religious grounds, but not only on those grounds. He believed that the parents whose taxes made schools possible should be able to have a say in what was taught in those schools. Bryan's majoritarian views put him on a collision course with the ACLU which championed the individual's right to free speech. Both parties have valid arguments. Surely both Bryan and the state of Tennessee were wrong to try to stop an idea either through legislation or through the courts, but we must respect their position that parents are entitled to some say in what is taught in a classroom. It is their tax money and their children.

Another reason Bryan opposed the teaching of Darwin's theory was his experience with its social applications. As a politician and public figure he opposed the business practices of men like Andrew Carnegie and John D. Rockefeller. He was the champion of those who bore the brunt of the hard business practices of these capitalists. Many capitalists justified their behavior in part by pointing to Darwin's teaching that, in nature only, the fittest survive and natural selection choses the favored races. Nineteenth-century capitalists saw themselves as the fittest survivors in the struggle for life. Bryan saw this as a great evil and one that was not supported by science since the theory lacked any empirical basis. For these reasons, he thought Darwin's theory should not be part of a public school education.

German militarism also played a role in Bryan's thinking. As secretary of state for Woodrow Wilson, Bryan promised there would be no

61 Larson, *Summer,* p. 136.

war while he held that post, so when the war came, he resigned. Bryan was aware that part of the rationale used by the Kaiser for the war came in part from Charles Darwin. Bryan's understanding was influenced by two scholarly books, *Headquarters Nights* by Stanford University zoologist Vernon Kellogg, and Benjamin Kidd's *The Science of Power.* Kellogg had gone to Europe and learned from German military leaders that natural selection served as a justification for the war; in fact, these men said the war would be "for the good of the world." Kidd's book noted how Friedrich Nietzsche was influenced by Darwin's theory of evolution and provided Nietzsche, to some degree, with scientific support for his radical ideas.

When Bryan got to Dayton, his reasons for challenging Darwin's theory in public school education were lost in his foolish response to Darrow's invitation to defend the Bible. The trial then became just a contest between Darrow's agnosticism and Bryan's fundamentalist Christianity. The foolish Bryan put himself in a position that no debater could have succeeded no matter how savvy. By allowing himself to be a witness on the stand, he was at Darrow's mercy since there would be no give-and-take, no debate. Years before, when commenting on how he would answer evolutionist claims, he had said he was prepared to do so, but after he responded to a question, he would be able to ask one of his opponents. In short, there would be give-and-take. It did not work that way in the trial and Bryan was humiliated.

Bryan was foolish to let himself be placed in such a position and wrong to defend countering Darwinian theory by legislation and/or litigation. That is a poor way to battle an idea. But the real story of the Monkey Trial is largely unknown. What people know or think they know is how an enlightened courageous lawyer hammered a narrow-minded doofus. The propagandists of 1955 did their work well, and the trial as portrayed in *Inherit the Wind* is the version most people carry in their heads. The irony today is that the roles of the play's hero and villain are now reversed. In the debate today it is the defenders of Darwin who work so hard like their fundamentalist opposites of yore to squash an idea, i.e., that Darwin may have not got it right and there may be a lot more to the development of nature than natural selection working overtime. Today it is Darwin's defenders who think they can stop an idea by having a judge make a ruling. Their tone is often more like the people of fictional Hillsboro than the real people of Dayton.

In their play Lawrence and Lee create the following exchange:

Drummond: "I object, I object, I object."

Brady: "On what grounds? Is it possible that something *is* holy to the celebrated agnostic?"

Drummond: "*Yes!* (His voice drops, intensely) The individual human mind. In a child's power to master the multiplication table there is more sanctity than in all your shouted 'Amens!', Holy, Holies, and Hosannahs!' An idea is a greater monument than a cathedral. And the advance of man's knowledge is more of a miracle than any sticks turned to snakes or the parting of waters!"[62]

At the beginning of the twenty-first century, the shoe is on the other foot. There are men of science today who have an idea and who are dedicated to advancing our knowledge of the natural world. Their challenge today is rooted in empirical science. They are skeptics challenging a nineteenth-century dogma. They maintain that while Darwin got some things right, he did not get it all right. It may be impossible for some people to ever come to terms with this. Like those who have seen Oliver Stone's *JFK* and "know" there was a conspiracy, there are those who have seen *Inherit the Wind* and "know" the story of the monkey trial. Their minds are made up and they do not want to deal with facts. But for those with an open mind we will see how this idea is fairing in the chapters to come.

---

62 Jerome Lawrence and Robert E. Lee, *Inherit the Wind*, Bantam Books, p. 93.

. . . the appearance of a new genus in the record is usually more abrupt than the appearance of a new species; the gaps involved are generally larger, that is, when a new genus appears in the record it is usually well separated morphologically from the most nearly similar other known genera. This phenomenon becomes more universal and more intense as the hierarchy of categories is ascended. Gaps among known species are sporadic and often small. Gaps among known orders, classes, and phyla are systematic and always large.

—George Gaylord Simpson

The known fossil record fails to document a single example of phyletic (gradual) evolution accomplishing a major morphologic transition and hence offers no evidence that the gradualistic model can be valid.

—Steven Stanley

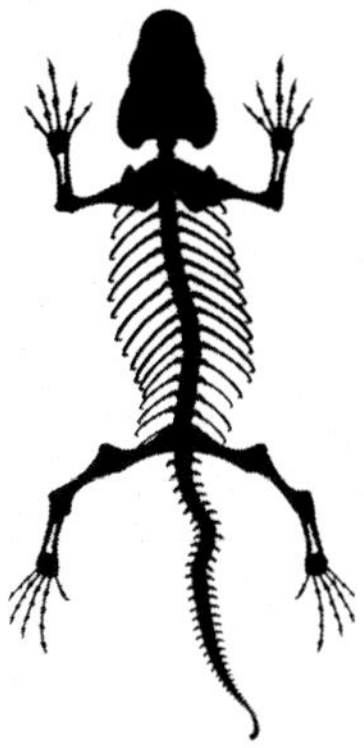

CHAPTER 10

THE FOSSIL RECORD

We Paleontologists have said that the history of life supports (the story of gradual adaptive change), all the while really knowing it does not.

—Niles Eldredge

"...one's imagination must fill up very wide gaps."

—C. Darwin in a letter to A. Gray

DARWIN'S ONE LONG ARGUMENT IN *Origin* relied heavily on the reader granting a grand extrapolation: what we do with domestic breeds, natural selection does with creatures in the wild. Darwin could point to changes in cattle, dogs, sheep and pigeons, but could not point to the great change he claimed natural selection had accomplished: crossing the species barrier. The changes made by natural selection were, according to Darwin, slight modifications and it took millions of years to get from one species to another. He was convinced such had happened and would never brook the idea that there were ever any jumps, *natura non facit saltum*. The evidence Darwin needed for this extraordinary claim could only be in one place: the fossil record. Only the fossil record could give a clue to what happened millions of years ago. The problem was the fossil record did not confirm the theory but contradicted the theory. This was no little problem, but as Darwin said, "the most obvious gravest objection which can be urged against my views."

The leading fossil expert in England, Richard Owen, rejected Darwin's theory in large part because he knew it could not be squared with the evidence. Georges Cuvier in France believed that the sudden appearance of new creatures and the sudden disappearance of creatures suggested that the earth had undergone catastrophes at different points in time. He theorized in this way because this conformed to the evidence in the fossil record. Not only do new creatures suddenly appear and disappear, but the basic characteristic of fossil record is stasis. That is, after a species appears in the record, it remains unchanged for millions of years. But according to the theory, because natural selection is working round the clock and creatures are in a nonstop struggle to survive, they are always evolving. You may recall Darwin had come to believe that the idea of species was outmoded and arbitrary because natural selection is always at work evolving living things. But there was a huge problem: the fossil record, as it was known then, did not in any way suggest that creatures evolved one from another. Darwin, greatly influenced by geologist Charles Lyell, believed that slow steady changes did the trick and imagined past catastrophes a shopworn idea in spite of the evidence of the fossils. (If Cuvier could read the accounts in today's papers of scientists discovering more and more evidence of meteor strikes during our earth's history, which they believe led to mass extinctions, he might be happily vindicated).

Here was Darwin's greatest dilemma. The only place that confirming evidence could possibly be found for one creature evolving step-by-step into another showed the opposite. Creatures appeared suddenly fully formed, remained unchanged for millions of years, then vanished. The intermediates that should precede them were not there. He himself said the transitional forms must be "innumerable" and there were none. Most troubling for Darwin was the record of fossils from the Cambrian era.

Approximately 550 million years ago in a very short amount of geologic time, about ten million years, the basic body types of the animals we know today appeared on the earth. The suddenness of their appearance was so great that it is common to refer to this part of the fossil record as the Cambrian explosion. Geologically speaking, the word explosion is well chosen. Nearly every major phylum that we know today appeared in this short time span. They appeared with no common ancestor and no intermediate forms to link them to the simple one-celled creatures that came before them. No wonder Darwin acknowledged this evidence as the "gravest objection urged against his views." The Cambrian explosion violates Darwin's theory not only because such varied and sophisticated life-forms seem to just pop up all of a sudden, but it violates his tree-of-life picture that we all carry in our heads. We all have been taught that the trunk of this tree consists of one or a few common ancestors, then variations cause our tree to branch out with greater and greater diversity with more and more life-forms. The Cambrian fossils show the opposite. No common ancestor, lots of life-forms to start with, and then the number of forms thins out over time. It is the reverse of the tree, or the tree turned upside down—the branches in the ground, the trunk up in the air.

What to say? In his defense Darwin said that due to different factors, the transitional forms were not preserved. No doubt many forms have not been preserved. But surely, if there had to be "innumerable" intermediates, the fossil record should show some and not just one or two. If there were good reasons why most forms were not preserved, would it not be fair to expect that even if ninety percent were lost, we could still see evidence from ten percent? Would it not be reasonable to expect, according to the theory, that the fossil record would be mostly intermediate forms, and that if we had only ten percent of those fossils available to us, we should have some sort of record of the intermediate creatures as one creature evolved into another.

As an example, we currently hear in the news about confirming evidence that birds evolved from reptiles. Given the extraordinary amount of body changes that must take place to go from reptile to bird, we should expect to find some of Darwin's "innumerable" forms. It will take a lot of transitional creatures to get us from reptile to bird. If we imaginatively try to picture the number of transitional forms needed to get from reptile to bird and take into account only four body changes—scales to feathers, solid bone to hollow, legs to wings and bellows lung to unidirectional lung—what is a reasonable number of forms needed to get the job done? (This sort of exercise was the kind of thing missing from *Origin* and is a major flaw in Darwin's work). Darwin and most of his defenders today hold that the changes along the way must be slight—no major jumps. Natural selection can only make slight improvements to a chance variation, oh and natural selection has no goal other than to provide an advantage for survival! Understand the feather itself is a stupendous feat of engineering that consists of hooks and barbules that lock and unlock. It must be and is very strong and very light. Remember also that the avian lung is unlike yours or that of an alligator. Our lungs are bellows lungs. Air flows in, then in reverse flows out. Not so the avian lung. It is unidirectional. Air flows in one direction, never reversing, so that the bird is supplied continually with needed air. And, as Michael Denton points out regarding the transition of a lung, air supply is absolutely vital for the creature to live. If our transitional forms experience trouble here, not enough air, our transitional forms are dead. Evolving is stopped right there. If this were not enough, the leg must change to a wing and the bones from solid to hollow. It is well perhaps to stop for a moment and quietly think about what it would mean for a reptile to evolve into a bird. Just close the book for a moment and imagine what an unguided process must accomplish just at the level of basic body parts.

Now let us continue with our imaginary number of intermediates as they go through nothing but "slight modifications." If it can be done in one million years, a small amount of geologic time, remember our process is unguided, and we allow for a life span of one hundred years for our transitional forms then we might have something like 10,000 forms. For whatever reason you want to postulate, ninety percent are lost to us; we can uncover only ten percent. We should have today in our museum 1,000 such forms. Ok, we can only find five percent, then we should have 500 forms. The museum display should show us the possible reptile ancestor that lived so many years ago at one end. At the other end of the display is the first bird, and in between 1,000 intermediates, ok 500 ...ok we can only find 100. But there is no such display! As it did for Charles Darwin, this lack of evidence for what the theory predicts should weigh heavily on anyone convinced his theory is an accurate picture of the origin of species.

Darwin's primary defense was the sketchy nature of the fossil record in his day. The work of unearthing fossils was just underway and most of that work confined to Europe, the United States, and Australia. He was convinced that, given time and as more fossils were found, the intermediate forms would be there. It was a fair defense in 1860. One hundred fifty some years later have his hopes been borne out? Do we have displays in our museums of present-day creatures and behind them the hundreds or thousands of forms that led up to them? We do not. In the years since Darwin made that defense a great deal of fossil work has been done, and at times done, primarily for the purpose of finding those forms and closing those pesky gaps. Even with this blinkered focus the evidence has not been found. The fossil record could never be a complete history of the possible changes a species undergoes, but it can be termed a fairly complete picture today of past life. Michael Denton in his *Evolution: A Theory in Crisis* provides a chart based on A. S. Romer's *Vertebrate Paleontology*, third edition. The following numbers may surprise many of us who think that the fossil record is only like snapshots taken here and there through history, giving us only a weak and sketchy record of the past.

Numbers of living orders of terrestrial vertebrates: forty-three; number of living orders of terrestrial vertebrates found as fossils: forty-two. The percentage fossilized is 97.7%. The numbers for living families of terrestrial vertebrates: 329; the number fossilized: 261, or 79.1%. If we exclude birds from the previous numbers, 178 terrestrial vertebrates; number fossilized: 156 for 87.8%. In conclusion, there is no reason for us to hold our breath at this point in time waiting for the fossil finds that will confirm Darwin's theory. We can say that the fossil record that was an enormous problem for him in 1860 is even a greater problem now. Today it is not true that we have only begun to see what fossils show us about the past. Decades of work around the globe have produced an enormous amount of evidence of past life-forms. Darwin's defense has been smashed by what has been found and by what has *not* been found.

Let us look back at the statements that begin this chapter. Except for the one from Darwin to Asa Gray, all the quotes are from twentieth-century paleontologists. All these scientists are committed to Darwin's theory in some form but know that the fossils are a problem. Why is it then, that most of us casual readers of science believe the opposite? We have all grown up led to believe that evolution happened in some Darwinian way because fossils testify to this. Why in this case do the experts know the truth and the public know a falsehood?

One of the reasons for this has to do with the birth of the theory. There were so many people, both the intellectuals of Darwin's day and just ordinary folk, who so wanted this new view of life to be true that clearheaded reasoning was at times hard to come by. Huxley and others helped set the tenor of the debate. The implications of Darwin's theory were enormous; the stakes were very high. In this atmosphere, cool rational give-and-take was lost in the battle to beat the opponent and win at all costs. In this atmosphere, the voices of men like Picet and Jenkin were lost. Their criticisms about the lack of evidence would not have been a death blow to the theory, but the problems they voiced were never answered. Subsequent years of fossil research has not answered them. The fact that the fossil evidence contradicted the theory was not minor quibble. Darwin's confident response that the evidence would be there after further research proved to be wrong. As the years rolled on without a satisfactory answer, it would have been fitting at some point for science to announce something like: "Houston, we have a problem"…"we need to reevaluate our theory in light of this lack of evidence." That, of course, never happened.

Not long after the publication of *Origin,* the fossil of Archaeopteryx was discovered. This famous ancient bird with reptilian features was immediately hailed as a link between species. Few believe that now, but after its discovery any new fossil find was simply said to confirm the theory. It was not possible in the climate that was established to find a fossil that contradicted the theory. A pattern was now in place that would carry into the twentieth century.

We might truly say that something like the emperors-new-clothes-syndrome was at work, and the needed honest appraisal never occurred. To say what the evidence truly showed would seem to put one at odds with modern science itself. Paleontologists saw stasis—that is, forms remaining unchanged for millions of years and not undergoing evolution. But like the emperor, that is *not* what they "saw." The following quote comes from *What Darwin Began: Modern Darwinian and Non-Darwinian Perspectives on Evolution* (Godfrey, ed., 1985):

> Each new generation, it seems, produces a few young paleontologists eager to document examples of evolutionary change in their fossils. The changes they have always looked for have, of course, been of the gradual, progressive sort. More often than not their efforts have gone unrewarded—their fossils, rather than exhibiting the expected pattern, just seem to persist virtually unchanged… This extraordinary conservatism looked, to the paleontologist keen on finding evolutionary change, as if no evolution occurred. Thus studies documenting conservative persistence rather than gradual evolutionary change were considered failures, and, more often than not, were not even published. Most paleontologists were aware of the stability, the lack of change we call stasis… But insofar as evolution itself is concerned, paleontologists usually saw stasis as 'no results' rather than as a contradiction of the prediction of gradual progressive evolutionary change. Gaps in the record continue (to this day) to be invoked as the prime reason why so few cases of gradual change are found.

Here is an example from the late twentieth century of the emperor's-new-clothes-syndrome. At the outset of this book I noted how I once believed that scientists were totally objective observers of our world, but I now know better. To some degree scientists in the last century have my sympathy. They have all had to operate in an environment that rewarded only one kind of finding and was ready to punish any results that strayed from the "faith."

Before we leave the subject of fossils we need to note two modern answers to those pesky gaps. One answer comes from Otto Schindewolf, who proposed that, at different points in time, evolution produced a new species in one jump. He proposed the kind of jump that Darwin would never countenance—a new species simply appears—voila! University of California at Berkeley's Richard Goldschmidt believed that a macro-mutation could at times achieve the same thing. Scientists have never given credence to this theory. Whatever the powers of natural selection or any other natural processes, the change involved is simply too great to be credible. A species is too complex and its many components too well adapted to make that kind of leap. If such a thing in any way did happen, it would be the same as a miracle. But sympathy is again in order when considering seriously this kind of saltationist evolution. A saltationist evolution fits the fossil record. Like it or not, Goldschmidt's and Schindewolf's version of evolution fits the evidence where Darwin's gradual change from one creature to another does not.

Another answer to the gaps problem is that evolution took place in some manner akin to Darwin's views, but the change takes place in a short burst of geologic time. Stephen J. Gould's and Niles Eldredge's idea of punctuated equilibra (or equilibrium) has been controversial since they first proposed it in 1972. Ernest Mayr, Gould's mentor, holds that there is really nothing in their view that differs much from what he had written before they created a stir with their now famous theory. Darwinists like Richard Dawkins who like their evolution one-hundred proof have had their disagreements with Gould and Eldredge. In Dawkins's *The Blind Watchmaker,* he takes a chapter to tell us that punctuated equilibria, while appearing to stray from the faith and water down Darwin to fifty proof, really does not when it is correctly understood. The issue Dawkins addresses in *The Blind Watchmaker* is the rate of change. If one's theory requires a rate of evolutionary change that is too fast, it then approaches the forbidden saltation. Dawkins says Gould and Eldredge appeared to sanction something like a saltation, but properly understood they did not. The rate of change, he assures us, lies within the required orthodoxy. We will not here examine in detail the debate generated by punctuated equilibria. We do, however, want to note one of its key provisions.

Gould and Eldredge say that evolution is not a process that is ongoing 24/7 effecting all the members of a species that are in competition for survival. (Darwin in later editions of *Origin* would say the same.) Rather, evolution occurs in small groups of creatures that get separated from the main group. Let us say such a small group journeys from the main body of creatures where they experience different environmental pressures and that a mountain range or some other natural barrier keeps them isolated. Due to environmental factors where they have migrated, this small group now evolves. Having evolved, they then return to the larger group and supplant them. The main group now appears to have evolved but the change took place in an area isolated from the main group. When we find fossils we find the fossils of the large evolved group, but not the fossils of the transitional animals that actually underwent evolutionary change. In this way, the fossil record shows what appears to be a sudden change without the transitional forms because those forms are lost to us. Gould and Eldredge's theory never gained much traction. It violated the gradual step-by-step model, and for that reason the fraternity dismissed it. What we can say without dispute is that the theory of punctuated equilibria is a theory for why we do not have evidence for evolutionary change. The convenient part of the theory is that it provides that evolution occurred in such a way that we cannot expect to find any confirming evidence. What those of us who doubt Darwin's theory need is not a theory for why we have no evidence, but evidence itself.

This has been a brief sketch of the problem that the fossil record has posed for Darwin's grand idea from the time of publication of *Origin* to present day. You will find a much fuller treatment in two excellent books: Michael Denton's *Evolution: A Theory in Crisis,* and Phillip E. Johnson's *Darwin on Trial.* One of the things that becomes apparent as one carefully examines the evidence that supports evolution is that the empirical evidence is confined mostly to small changes within species. This kind of change was well known in Darwin's day. The only place where the grand claim that all life evolved from one or a few common ancestors by natural selection and/or other factors has not been confirmed by the record we have of past life on the planet. In the July 17, 1981 issue of *Science Magazine*, David M. Raup, then with the Field Museum of Natural History in Chicago, wrote a letter to *Science Magazine*. He wrote, he said at the outset of his letter, to correct what he termed "misinformation" about the fossil record. The following quote comes from the third paragraph of his letter:

> A large number of well-trained scientists outside of evolutionary biology and paleontology have unfortunately gotten the idea that the fossil record is far more Darwinian than it is. This probably comes from the oversimplification inevitable in secondary sources: low-level textbooks, semipopular articles, and so on. Also, there is probably some wishful thinking involved. In the years after Darwin, his advocates hoped to find predictable progressions. In general, these have not been found—yet the optimism has died hard, and some pure fantasy has crept into textbooks.[63]

63 *Science*, July 17, 1981, Vol. 209, No. 4505.

CHAPTER 11

LESSONS FROM MOLECULAR BIOLOGY

Can all of life be fit into Darwin's theory of evolution? Because the popular media likes to publish exciting stories, and because some scientists enjoy speculation about how far their discoveries might go it has been difficult for the public to separate fact from conjecture… if you search the scientific literature on evolution, and if you focus your search on the question of how molecular machines—the basis of life—developed, you find an eerie silence. The complexity of life's foundation has paralyzed science's attempt to account for it; molecular machines raise an as-yet impenetrable barrier to Darwinism's universal reach.

—From *Darwin's Black Box,* by Michael Behe.

WE HAVE BEEN NOTING WHAT a stumbling block the fossil record is for anyone wishing to place their confidence in Darwin's theory of evolution. The fossil record shows simple life-forms appearing first in the oldest uncovered strata and then more complex life-forms appearing later in more recent strata. The progression from simple to complex in the record is evidence enough for many that Darwin was right. They believe the fossil record is an asset. Others, and we have looked briefly at the views of some of Darwin's contemporaries, do not find it convincing. All agree that the record is one of simple life-forms appearing before the more complex ones, but critics do not find this evidence convincing. For them, to have confidence that from one or a few common ancestors all life-forms have evolved from a totally unguided process, the sequence of simple to complex is not sufficient. The great body changes that must have occurred, the time needed for great and small changes, the apparent design and purpose of much of nature and the lack of evidence of one species morphing into another, prevent them from giving assent to Darwin's proposition.

This is and always has been the nub of the issue about Darwin's theory. How much evidence does it take for a person to find the theory compelling? Or, to look at this question from the other angle: though the theory is plausible, for skeptics there are too many hurdles, and so the theory fails to convince. Darwin's critics were not unfair. They granted that his book had much merit and that his ideas were well presented, but there was not enough evidence to overcome the many problems. Hurdles that Darwin thought minor and ones that he believed he had answered along the way, they believed were large and unanswered. Today there are a growing number of skeptics who believe the hurdles are getting higher and higher. They find believers in Darwin too easily satisfied.

We want to bring our examination of the theory of evolution into the present day. We do so understanding that we are not going to disprove Darwin's theory any more than someone else can prove it. It is not a thing to be proved. How can it be possible, even in theory, to prove a theory that deals with events hundreds of millions of years ago? Those who believe Darwin's theory to be as proved as the idea of the rotation of the earth are simply talking nonsense. Evaluating events of millions of years ago will always be a matter of judging the evidence; what one will find convincing, another will not.

Let us examine the problems raised by modern science and made possible by modern technology. We will look at the world of the cell, the world of astrophysics and the world of information. This, I believe, will only reinforce doubts. For those who have at any time wondered if there were good reasons to question whether Darwin explained it all in 1859, it is hoped that what follows will encourage the skeptic in you. Others will weigh the scales of evidence and find modern objections to evolution too light to tip them into unbelief. It is hoped that those who are Darwinists of one stripe or another will respect those of us who find the theory wanting. We ought to always disagree respectfully.

In the last chapter we took a moment to imagine what would be involved for a reptile to evolve into a bird. I would like to revisit that idea. Georges Cuvier, Darwin's French contemporary, believed that species were immutable, i.e., they could not evolve from one kind of creature to another. One of the reasons he and many others took this position is the extraordinary combination of complex and interrelated systems composing a creature. The scales of a reptile serve a function that makes it successful in its environment; Darwin's theory tells us they are slowly going to change into feathers. A feather is something very different from a scale. A great number of transitional creatures would have to survive for some period of time without true scales and without true feathers. Yet natural selection is said to confirm an advantage for survival. How can a transitioning creature be better suited to survive when it lacks true scales but does not yet have feathers? This sounds to the skeptic much more like a disadvantage than an advantage.

Scales are just one part of the many interacting systems that make up a reptile and that must evolve in a coordinated way with all the other systems. While the scales are changing, the bones must also be changing. What good will the feathers do if the creature is too heavy to fly? When one factors in such complexities, it becomes clearer why scientists in Darwin's day thought he was on the wrong track. If a leg is evolving into a wing it must at some point be a poor leg and a handicap for survival. But it is not yet a wing that can create flight. If the bones have not grown hollow when the wing appears the creature will be too heavy for flight. Again, where is the advantage for survival? We also must keep in mind that, according to Darwin and his modern-day supporters, natural selection has no goal in mind when scales turn into feathers. There is no natural process trying to create a bird. There is no purpose other than survival. Given what must have been tremendous disadvantages along the way, how did the transitioning creatures survive?

We must also ask what kind of environmental factors would be necessary to move the creature steadily over time along the path from reptile to bird? For thousands or millions of years the environment would have to be just so to enable all of the many systems of the reptile to be pushed to an avian form. All of this points to some of the most fundamental problems that must be overcome to explain the change in the main body plans. These problems are daunting.

But if these major body changes were not enough to challenge the theory, modern science has compounded the problem of creatures evolving many times over, thus deepening the case for skepticism. Thanks to modern technology, in the study of the cell we know a great deal more now than Darwin did one hundred and fifty years ago. For Darwin, the cell was simply a very small undifferentiated blob. Its true nature had not yet been revealed. There was no way for Darwin with his rather crude scope to see what scientists can see today. That brings us the challenge posed by the micro world of the cell—specifically the challenge posed by Dr. Michael Behe in his book *Darwin's Black Box.* But before we have a look at the nature of this challenge, let us consider a biographic issue.

From his youth, Darwin was encouraged to believe in an evolving nature. It was part of his family heritage. It was fostered in Edinburgh. Events, especially the episode with the Fuegian missionaries, moved him to connect all creatures in one related family. Back in London, his circle of friends continued to reinforce his evolutionary thinking. When he began his notebooks in 1838 there was little left of his Christian faith, which died with the death of his beloved Annie. All his life was a journey to a conviction of what his mentor, Edmond Grant in Edinburgh, exposed him to as a teenager: species evolve by strictly natural processes with no need of an intervening God. The result of this mental and philosophical journey was his great book.

It is often said that today's critics of Darwin are nothing more than an updated version of the Bible-thumping Matthew Harrison Brady—just slightly more sophisticated. They are said to come from the same stock, but this is simply not the case. Dr. Behe is one of Darwin's most pointed critics but his journey into doubt is the opposite of what many might assume and is a contrast to Darwin's own journey. Behe, like you and me, was raised and brought up to believe in Darwinian evolution. Charles Darwin let the creationist view of the world slowly slip away as he moved more and more toward evolution. His was largely an imaginative journey that required just enough evidence found in minor observable changes within species to justify him becoming an evolutionist.

Behe was educated in Catholic schools, then in a secular university setting. In all this schooling, he was taught that Darwin's evolution was the truth of biology. The basis for doubts would come later and not from philosophy or theology or economic theory, but from the hard facts of molecular biology. His schooling was not a journey of doubting Darwin, but of believing in Darwin. In an article in the journal *First Things,* Behe related how his wife, also given a Catholic education, was supposed to be startled by her science teacher proclaiming to good Catholic children, "Evolution is true. Get used to it."

He tells us the striking thing about this event was that it was not at all striking. His wife and her classmates "shrugged." When they were told that God could do his creating by secondary causes like natural selection, this only showed God's great power—"and who were they to tell God otherwise?" They were taught that God put creation in motion in such a way that over time natural processes could create all that was wanted by the creator including human beings. In short, they were taught that evolution was the way God made the world. Behe and his classmates were not in the least troubled by this view of creation. Later in life in secular universities, when the issue of the origin of life was presented as something again from natural processes only, he had no trouble accepting the idea because he tells us his Catholic education prepared him for just that conclusion.

He shares a revealing story from his time of postdoctoral studies at the National Institute of Health in the early 1980s: "I shared a lab with a woman named Joanne, a fellow postdoc and a serious Catholic. One slow afternoon she and I were gabbing about the Big Questions, including the origin of life. 'What would be needed to get the first cell?' she asked. 'You'd need a membrane for sure,' I said. 'And metabolism,' Can't do without a genetic code,' she added, 'and proteins. Naaaaah!' Then we laughed and got back to work. Even though we realized there were brick walls everywhere one looked, our only reaction was to chuckle. What we didn't do was to question seriously whether the unfolding of physical laws could adequately explain the very start of life. I guess we vaguely thought that even if we didn't know, somebody else must."[64]

Who among us cannot relate to this little exchange? Most of us have never been postdoctoral scientists, but we can relate to questioning how natural processes alone might account for all our experiences and all the physical universe. The doubt is perhaps unnamed but felt like a burr. But we let our doubt slide on by and like them assume "they" have accounted for it… somehow. The fact that his story is about two postdoctoral scientists demonstrates the strength and pervasiveness of the Darwinian worldview. We have all grown up with this worldview that governs yet today. Its power is demonstrated by how it overrides our doubts.

How then did a man trained in Catholic schools and secular universities to believe in Darwin come to think otherwise? Surprisingly, or not surprisingly, what set Dr. Behe on the road of doubt was his professional life's work: the world of the cell. It was the empirical world. Hard facts moved Behe from casual acquiescence to Darwin's theory to doubting it. The result is his book, *Darwin's Black Box.*

Behe's pointed critique of how a gradual step-by-step process could produce molecular machines drew serious reviews from many publications. The book's pointed focus opens, even to lay people, a micro world of wonder. Darwin's critics in 1860 were just in their criticisms of his theory given the world of science they knew. What Dr. Behe has done in *Darwin's Black Box* is to bring that same questioning spirit to the world of biochemistry. He is a fine and fun guide. His book opens up for us a dazzling micro universe that is much more complicated than we might ever have imagined, and one that fills us with deep and true awe. Most of us have little knowledge of the cell. He helps us to see new hurdles that Darwinism must convincingly overcome if we are to have confidence in it.

First, what is a "black box?" Behe uses the term to mean a device whose inner workings are a mystery. For Darwin and his contemporaries, the workings of the cell comprised a black box. No one in his day had a clue how the cell did what it did. But with the advent of the electron microscope the black box of the cell began to be opened. It is Behe's contention that while Darwin's natural selection may explain

64 *First Things*, December, 2005, No. 158.

somc aspects of nature, it cannot adequately account for the origin of complex biochemical systems. He further contends that if Darwinian theory cannot adequately account for the goings on in the cell it is in serious trouble. His book opens for us Darwin's black box and asks us to consider the difficulties Darwinian theory faces in trying to account for the workings of life at the molecular level.

In five chapters he has us examine molecular machines, the bacterial flagellum, the cascade that is our blood clotting mechanism, the highways of cellular transport, our defensive immune system, and lastly the molecule AMP. He explains in some detail how these things work so that we can gain an appreciation of how difficult it is to account for the origins of life at the molecular level with only unguided natural processes.

We will begin our look at biochemical systems with the bacterial flagellum. But first we need to consider the term "irreducible complexity"—now part of the current debate about evolution thanks to Dr. Behe. It is his contention that, at the biochemical level, there are many systems that exhibit this characteristic and that such systems pose a special problem for Darwinian development. This is how he explains irreducible complexity in *Darwin's Black Box*:

> "...a single system composed of several well-matched, interacting parts that contribute to the basic function, wherein the removal of any one of the parts causes the system to effectively cease functioning. An irreducibly complex system cannot be produced directly (that is, by continuously improving the initial function, which continues to work by the same mechanism) by slight, successive modifications of a precursor system, because any precursor to an irreducibly complex system that is missing a part is by definition nonfunctional. An irreducibly complex biological system, if there is such a thing, would be a powerful challenge to Darwinian evolution. Since natural selection can only choose systems that are already working, then if a biological system cannot be produced gradually it would have to arise as an integrated unit, in one fell swoop, for natural selection to have anything to act on."[65]

---

65 Behe, Michael, *Darwin's Black Box*, Touchstone, New York, 1996, p. 39.

As an example of a system that is irreducibly complex he presents a simple mouse trap: wooden base, spring, catch, holding bar, and hammer. All parts are required for the mouse trap to function. You cannot start with one, two or three parts, catch a few mice, and with slight modifications build up to the complete system that catches lots of mice. All parts have to be present for the trap to work, and the parts must be matched well enough to meet a minimal function. The spring must be strong enough to stop the mouse but not too strong or the catch will not work. The catch must hold the hammer lightly enough for the mouse to trip it and so on.

When molecular biologists refer to the tiny systems that perform different functions, they use the word machines—molecular machines. They do so because the term molecular machine is a perfect fit for these micro systems. Our lives are filled with machines that perform work for us. Under the hood of our cars is a machine built to deliver power to take us across town. The machine that produces power is connected to a system we call the transmission that delivers the power to the wheels. All the systems together make up our automobile. On a grander scale our bodies are made up of many machines that are linked together in biological harmony. Those who defend Darwin's view of life believe that your body is the result of unguided natural processes that, over time, have been able to produce everything from the simple cell organisms to creatures as complicated as yourself. Dr. Behe asks us to consider the challenges that unguided natural processes would face trying to create the necessary molecular machines.

He first has us consider the cilium and the bacterial flagellum, both used for swimming. The cilium works like a paddle to move a cell through its liquid environment. The bacterial flagellum moves a cell like a propeller moves a boat. Both of these swimming machines meet Dr. Behe's definition of irreducible complexity: that is, each is a system of well-matched parts wherein all of the parts must be present for the machine to function.

Like the outboard motor on your boat, the bacterial flagellum is comprised of a stator, rotor, drive rod, bushing, universal joint, hook, and the propeller. The components are made of proteins rather than metal. (The proteins themselves are comprised of amino acids, but we will not take time to examine how amino acids come together to make proteins. As one explores this level of biological life, the complexity just continues at lower and lower levels). Some twenty to twenty-four proteins make up the flagellum that can rotate the propeller at speeds faster than formula one race cars, up to 100,000 rpm. The propeller can move in both clockwise and counterclockwise directions.

Not surprisingly, biologists have been fascinated by this machine since its discovery in 1974. It has been the subject of research for over thirty years now. The assumption among most biologists is that somehow this machine evolved in Darwinian fashion. But after more than thirty years, research has yet to present a single concrete step by molecular scenario of how this machine might have evolved. Given the nature of an irreducibly complex system, this is not so surprising. A machine of this caliber presents a special challenge to Darwinian evolution. Since Dr. Behe's book, the bacterial flagellum has been subject to greater scientific scrutiny. One of the Darwin's defenders, Kenneth Miller Professor Emeritus of Brown University, has proposed the type three secretory system (TTSS) of some bacteria as an answer to this challenge. There are components in the TTSS system similar to the base of the flagellum's motor. Given the similar function, Miller believes that the motor could evolve from the TTSS. He believes it shows that the flagellum is not irreducibly complex and therefore a thing that could have evolved. If some parts of the machine are removed, there are parts remaining that still function. Here is a brief response from Dr. Behe:

> In recent years it has been shown that the bacterial flagellum is an even more sophisticated system than had been thought. Not only does it act as a rotary propulsion device, it also contains within itself an elegant mechanism to transport the proteins that make up the outer portion of the machine, from the inside of the cell to the outside. (Aizawa 1996) Without blinking Miller asserted that the flagellum is not irreducibly complex because some proteins of the flagellum could be missing and the remainder could still transport proteins, perhaps independently. (Proteins similar—but not identical—to some found in the flagellum occur in the type III secretory system of some bacteria. See Hueck 1998). Again he was equivocating, switching the focus from the function of

> the system to act as a rotary propulsion machine to the ability of a subset of the system to transport proteins across a membrane. However, taking away the parts of the flagellum certainly destroys the ability of the system to act as a rotary propulsion machine, as I have argued. Thus, contra Miller, the flagellum is indeed irreducibly complex. What's more, the function of transporting proteins has little directly to do with the function of rotary propulsion as a toothpick has to do with a mousetrap. So, discovering the supportive function of transporting proteins tells us nothing about how Darwinian processes might have put together a rotary propulsion machine.[66]

Whether you find Miller's answer to the challenge convincing or not, it is at best a hypothetical supposition. The truth is, and it is very important to keep before our minds, that no one *knows* how the bacterial flagellum came to be. Thirty years after the discovery of a machine capable of turning 100,000 rpms, science does not *know* how evolution could produce such a wonder. (For fuller airing of Dr. Behe's response to his critics see his book *A Mousetrap for Darwin: Michael J. Behe Answers His Critics*).

Following the look at the bacterial flagellum, Behe shares with us the wonders of the blood clotting cascade. He begins the chapter with a look at Rube Goldberg cartoons depicting Rube Goldberg contraptions. For those too young to know, a Rube Goldberg cartoon would present a "machine" with a series of interacting steps—the first triggering the second, the second the third, and so on until, at the end of the sequence, the needed result is achieved. Some may remember from their childhood a board game called Mousetrap that used a Rube Goldberg kind of system. In this game, the players assemble a series of linked parts so that, at the end of the game, one could set the cascade in motion to trap a plastic mouse. When all the parts were on the board and connected to one another, a player turned a crank to set the whole sequence in motion that involved rubber bands, a marble, ramps, etc. Once the initial step was taken, if all went well, each part triggered the next until, at the end, a cage fell from its perch and the mouse was trapped.

66 From *Still Spinning Just Fine: A Response to Kenneth Miller*, by William A. Dembski, 2.17.03, v.1.01.

When you or I cut our finger while preparing the dinner salad, there follows a sequence of that nature that stops your finger from bleeding. The sequence is called a cascade because one step triggers another, and then another. If we remain for a moment with the Mousetrap game in our minds, we can immediately sense how such a sequential system challenges Darwinian evolution. Each link is vital to success of the needed outcome, and all must be present together to work at all. It is very difficult to conceive of how this could come to be in a step-by-step process. Here we will not take the time and space to share the entire cascade as described by Behe. However, it takes about five pages for him to cover the steps in the cascade, and these multiple steps make Rube Goldberg look simple by comparison. Let me quote from just *one* part of the cascade describing the extrinsic pathway for just *one* of the components called Stuart factor. Remember, the following paragraph describes just one section of the total blood clotting cascade.

> "When an animal is cut, a protein called Hageman factor sticks to the surface of the cells near the wound. Bound Hageman factor is then cleaved by a protein called HMK to yield active Hageman factor. Immediately the active Hageman factor converts another protein, called prekalli-krein, to its active form, kallikrein. Kallikrein helps HMK speed up the conversion of more Hageman factor to its active form. Activated Hageman factor and HMK then together transform another protein called PTA, to its active form. Activated PTA in turn, together with the activated form of another protein called convertin, switch a protein called Christmas factor to its active form. Finally, activated Christmas factor, together with antihempohilic factor (which itself activated by thrombin in a manner similar to that of proaccelerin) changes Stuart factor to its active form."[67]

From this sample of what the cascade involves you know that Rube Goldberg or the Mousetrap game hardly begin to touch on the staggering complexity of how your blood clots. Add to this sequence the issue of regulation. Your blood has the capacity to clot so that you do not bleed to death from a cut. The cascade must not start unless it is needed. That is, it

---

67 Behe, Michael, *Darwin's Black Box*, Touchstone, New York, 1996, p. 84.

has to be turned on at the correct moment and then begin to work only at the wound. Furthermore, at the correct moment the clotting must stop. If it is not turned off at the correct moment, if the clotting process goes too far, the rest of your blood will be in danger. If a cut triggers clotting, but the clotting process is without sophisticated regulation, from a simple cut on the finger you could be dead. The need for highly sophisticated regulation is not a luxury for the system, but a matter of life and death.

The question Behe wants us to have vividly before our minds is: how could this cascade arise in step-by-step fashion when one or even a few steps would be useless? Less than all necessary steps spells failure. How could an unguided process have a first but useless step, and from there build on that year after year to create a functional cascade? Furthermore, at any time along the way of evolving, if the cascade is not adequately regulated, the evolving creature dies. It either bleeds to death or dies from too much clotting.

The remaining examples from *Darwin's Black Box* are no less challenging than the two we have just covered. Vesicular transport, the immune system, and the molecule AMP add hurdle upon hurdle for the theory of evolution. What is detailed here are machines at the molecular level. These machines have to be integrated with others in cooperation for the complete organism to function. Behe's conclusion is that since science does not know how these systems came to be formed by a step-by-step process, and since there are inherent theoretical problems imagining how it might have been done, there must be a place for design by an intelligent agent to help account for it. At this point you may think that Behe or any who are sympathetic to his conclusion have abandoned science. If so, you have my sympathy. Given our educational system and the way science and evolution are presented to us today, it is a natural response. For some of us it is the only response we are capable of. But I ask for your patience. I am asking for an open mind willing to consider an alternative point of view. For many, just the mention of the word design throws up red flags, but I am asking you to hold your fire for a few more pages.

CHAPTER 12

THE POSSIBILITY OF DESIGN

The theory of intelligent design holds that certain features of the universe and living things are best explained by an intelligent cause, not an undirected process such as natural selection.

—definition from website of the Discovery Institute.

AT THIS POINT IT WOULD be well to say a few words about intelligent design—a hot topic to be sure. We will have more to say later but since Behe's book takes us there, let's look at this idea for a moment. You may have read that the argument is fundamentally a religious one, not a scientific one. If you read the writings of those who champion intelligent design, you will know that the charge is not true. We know in Behe's case this is not true. His religious education taught him to believe in Darwin's theory of evolution; it was his experience as a scientist that caused him to doubt what he had been taught. His doubts do not spring from Christianity or a reading of the book of Genesis, but from empirical study. The reasons for his doubts are crucial. Behe, like so many other skeptics of Darwin, was once like all of us, a believer in Darwin's theory. It was the hard facts of science that compelled him, as it has so many others, to think otherwise.

Just as intelligent design does not begin with religion, it does not end there. An agnostic could readily acknowledge design in the world of the cell. Those conceding the evidence for design are not making claims of who the designer is or ascribing motives. The detection of design does not prove God exists or anything like the truth of Christianity or any other religion. (Frances Crick famously said that life and design on planet earth might have been the result of visiting aliens from outer space). Detecting design is an empirical endeavor—very scientific. Our own experience tells us that intelligent agents can and *do* produce designed things. We *know* an intelligent agent can produce something designed. We are surrounded by design—the office you work in, the car you drive and the computer you use are all products of intelligence. The opposite is also true. We do not have the experience of unguided natural process producing complex machines. We have no experience in our daily life of random processes producing design. In fact, there is an enormous *a priori* improbability of such a thing. Even the humble mouse trap, simple as it is, is something we recognize as a thing which requires an intelligent agent.

If we are walking in the woods and come upon a crude fence made of logs or stones, we immediately know that nature did not do this on her own. If this *a priori* improbability of randomness and chance producing design applies with such force to the crude fence in the woods, is there anything out of line in bringing this same reasoning to the molecular structures described in *Darwin's Black Box?* Certainly the systems described by Behe are much more complex than the crude fence. In the absence of a Darwinian explanation for molecular machines, not guesses as to how they might have been built, does not reason herself require that, at some point, we must be dealing with design to some degree?

Consider that we are familiar with design in everyday life, and that it is not necessary that the designer be known. We are all familiar with the TV crime show. A body has been discovered and the perpetrator is at pains to conceal the deed. The murder victim does not show signs of violence, yet the good guys suspect foul play. A pathologist examining body tissues is able to detect poison, and so the good guys know that this death was designed, and we know the victim was murdered. The good guys may not find the murderer and he or she may remain unknown. Even though the designer of the death remains unknown, there is no doubt from the evidence that the victim's death was result of a malevolent intelligence.

Another example of science working to detect design can be found in Carl Sagan's story *Contact.* Sagan's novel is based on the real scientific project called SETI—search for extraterrestrial intelligence. The project uses real science; the premise is that design can be detected. If you did not read the book, you may have seen the film made in 1997 starring Jodie Foster and Matthew McConaughey and directed by Robert Zemeckis. In the story, radio astronomers monitor radio signals from outer space searching for patterns that would indicate intelligence—a pattern that is not random but specified. As long as the astronomers hear nothing but random patterns there is no reason to believe there is anybody, any intelligence, trying to communicate with earthlings. Random signals are what nature does on her own, just as waves leave patterns of shells on a beach. But what would a beach-walker think if he crossed a headland and found the shells on the beach ordered in a way that spelled, "John loves Mary"? The person walking on the beach knows that the random processes in nature do not produce a pattern with shells that convey a message. This is without doubt a product of intelligence. In Sagan's story the radio astronomers discover a pattern of beats and pauses that correspond to the sequence of all the prime numbers between two and one-hundred and one. Now they have a sign, because the pattern is not random, that some form of intelligence in outer space is communicating with us. The detection of design in *Contact* is thoroughly scientific and thoroughly empirical just as it is in Behe's *Darwin's Black Box.*

We can see from these two examples that detecting design is a logical inference. It is an experience common to us all and is based on reasoning and noting the best possible answer to how certain things came to be. The question we wrestle with today comes right out of *Inherit the Wind.* Will scientists, and the rest of us, be allowed to entertain such thoughts? Is it the business of science today to attack such thinking because it challenges Darwin or because it has different implications than his theory did? We know that part of the attack on intelligent design comes from those who are upset with the implications that point to something beyond nature. That is understandable. Some of those who attack Darwin do so only because of the atheistic implications of this theory. Both ideas say nothing explicitly about religion, but the subject is close at hand. If one has implications, then both do. As far as the debate regarding the merits of one or the other, the implications can balance one another. The debate must be based on the merits of each idea, not on the implications for each idea.

It is sometimes said that to attribute design to aspects of nature is giving up on science. But the scientists who argue for design are doing nothing of the sort. How could coming to the best possible conclusion following the evidence be giving up on science? In some cases, it may mean clearing the deck of an outmoded or unproductive theory in order to get on with the work of studying the natural world.

Finally, some say that science cannot proceed if it goes down this road. To acknowledge design, it is said, will bring science to a halt. But if such was the case, then no science could have occurred prior to Darwin's theory. It would be as if scientists could only stumble in darkness for centuries until they were liberated by Darwin's theory. How was it possible for a Newton to do science if he first had to wait for Darwinian liberation? All one needs to do is to reflect for a moment upon the Western scientific enterprise which was born out of Christianity. Men like Copernicus, Galileo, Bacon, Newton, Pascal, and Faraday were all Christians of some stripe. Even those men of science who would not call themselves Christians operated in a world in which God was thought to be the author of the universe. Until Darwin, even the scientist who would not call himself a Christian worked in a Christian framework. Much was accomplished in those years.

"Biology is the study of complicated things that give the appearance of having been designed for a purpose,"[68] writes Richard Dawkins in the first chapter of his defense of Darwinism in *The Blind Watchmaker*. There is no question, he tells us, that what we see in nature looks like it was designed, but in the next three hundred pages he tells us why things only *appear* to be designed. We have been fooled. Design is only an illusion he tells us. His book is devoted to showing how natural selection has done the work of creating all things that look designed but are not, and thus a designer is unnecessary. Dawkins has absolute faith in natural selection to explain all, even though we have seen from the get-go how little empirical evidence there is that it can accomplish anything but minor changes within a species. Dawkins is so committed to a world

68 Richard `Dawkins, *The Blind Watchmaker*, W.W. Norton Co, New York, 1996, p. 1.

without any kind of designer that his faith in natural selection is unshakable—the evidence be damned.

Perhaps it is time to re-examine whether these "complicated things that give the appearance of having been designed for a purpose" are just that—complicated things that have been designed for a purpose. Is it possible that nature is *not* fooling us, but speaking truthfully? Is it possible that when we see design we are not being fooled? Is it possible that our first instinct is the right one? Is it possible that when we see design and see the obvious, we see truly?

Dawkins's *The Blind Watchmaker* is long on theory and short on evidence. Those who are like my neighbors in Missouri, the "Show-Me" state, will not find much to sink their teeth into. Dawkins spins quite a story about computer generated bio-morphs while dismissing all who are skeptical of natural selection's creative powers. Today the books defending the design argument are steadily increasing in number and in scope, and the inquisitive reader will want to read from both sides of this issue before coming to his or her own conclusions. This battle of ideas is just recently joined and makes for very exciting reading.

But we will close this brief look at the issue of design in nature with a simple and uncomplicated thought: if it looks like a duck, waddles like a duck, and quacks like a duck, then maybe it is a duck.

CHAPTER 13

LESSONS FROM OUTER SPACE

Many proponents of accident or chance as the cause of life proposed their theories when the universe was considered by many scientists to be in a steady state with infinite age. Recent discoveries in modern physics and mathematical analysis accompanying those discoveries, however, changed this view of the universe to that of a young universe, expanding from a definite beginning. From this perspective the probability calculations concerning the precision of many of the values in particle astrophysics present a strong case against an accidental universe.

—From *A Case Against Accident and Self-Organization,* by Dean L. Overman.

LET US SHIFT OUR FOCUS for a moment. We have been considering the world that is visible to us through the electron microscope. Let's trade in our microscope for a telescope and consider for a moment the world of astrophysics and outer space. This change of focus will provide a contrast in subject matter, and more importantly, show the contrasts in the responses of scientists regarding intelligent design.

In 1929, Edwin Hubble, working at the Mount Wilson Observatory in Pasadena, California, discovered that the universe was expanding. Hubble noted that galaxies were moving away from one another at a velocity directly proportional to their distances apart. This discovery meant that the universe was expanding. If the universe is expanding, then it must have a starting point. This starting point is called a singularity.

This idea of a definite beginning was inimical to a great many scientists. The popular term "Big Bang" was at times used derisively by those who did not like the idea that the universe had a definite beginning, as in "do you mean it all started with a Big Bang?" Well, yes, in a sense it did. We are not talking about an explosion, but an event that is nearly indescribable. It is not only the beginning of matter and energy but of space itself. The idea, cherished by so many, that the universe existed in some steady state, had to be abandoned. The depth of hostility to the idea of a definite beginning for the universe can be judged by the fact that Einstein's general theory of relativity showed an expanding universe. But even Einstein was not immune from the pressure to show a steady state universe. He therefore altered his equation to meet the demand that the universe did not have a beginning. This came to be known as the "fudge factor" and perhaps the greatest mistake in the great scientist's life. He was right the first time, and Hubble's observations confirmed it. Today most scientists accept the Big Bang and place that event at somewhere in the neighborhood of fifteen billion years ago.

As physicists and astrophysicists have explored how we got from that singularity to the present, they have discovered that it could only have happened if certain conditions were met along the way. The words "certain conditions met" hardly scratches the surface of what they discovered. The forces and the elements of the universe have had to have been calibrated to nearly unbelievable tolerances. In any one of the following examples, if forces and dimensions were not "just so" the universe would not exist as we know it. This fine-tuning of the universe adds another dimension of doubt regarding how unguided processes alone could account for our world.

Before we begin we should note that we will be dealing with very large and very small numbers. They will be expressed in powers of ten using exponents. For example, 10^4 is a way of representing 10x10x10x10 or 10,000. The number for one billion would be expressed 10^9, and 10^{10} is 10 billion.

With the Big Bang the universe is in play and expanding and at just the correct rate of expansion. As it expands the force of gravity balances the force of expansion at just the right level. If the universe expands with slightly greater force than it should, stars, galaxies and planets will not form. The matter of the universe would simply scatter like sand and not coalesce. Matter must coalesce in order for stars and planets to come into existence. If the force of gravity was just slightly greater than it is, then gravity would eventually halt the expansion of matter altogether. After so many millions of years, the universe would collapse back on itself. The universe would have experienced a "big crunch." The matching of these two forces had to be to the remarkable precision of one part in 10^{55}. Working with factors such as the background radiation in the universe leftover from the Big Bang, the age of stars, and the expansion rate of the universe, scientists now calculate that our universe is some 13.7 billion years old, give or take a billion. Yet physicists can speak to the crucial balancing they believe must have occurred within the very first second of the universe. Imagine if you can fifteen billion years. Now consider that certain factors for the success of the whole enterprise are crucially balanced within the very first second of the fifteen billion years.

Planck time is the moment when the cosmic clock begins to tick. In the first second after Planck time, again the force of expansion and the force of gravity have to be balanced perfectly. The density and the critical density of the universe had to be balanced perfectly. It is believed that all four of these elements were properly balanced in a very short period of time after Planck time. This period of time is known as the inflationary epoch. The inflationary epoch began at 10^{-36} of the first second of the Big Bang and ended at 10^{-32} of the first second of the Big Bang, an inconceivably small amount of time.

When considering a moment in time this small it is helpful to have a picture. Cosmologist Paul Davies of Arizona State University provides help with this illustration. You are going to fire a bullet at a target one-inch square—think the end of your thumb. Now for the distance from you to your target: your one-inch target is on the other side of the observable universe, which is twenty billion light years away. The odds of you hitting that target are one in 10^{60}. Those are the odds of the four forces being balanced during the inflationary epoch.

There are four fundamental forces of the universe: gravity, the strong nuclear force, the weak nuclear force, and electromagnetism. "The strong nuclear force binds the particles in an atom's nucleus and is the strongest of the forces, approximately one hundred times as strong as the electromagnetic force, which in turn is ten thousand times stronger than the weak nuclear force. The weak nuclear force is approximately ten thousand billion, billion, billion times stronger than the force of gravity. Considering these enormous differences in strength, one can appreciate the remarkable precision required to balance these forces to a degree of one part in 10^{60}."[1]

Again and again we learn that the odds of having things balanced to the correct degree needed are beyond belief. Atoms are made up of electrons and protons that must number in the proper ratio. The number of electrons in our universe must be balanced with the number of protons to a degree of accuracy of one part in 10^{37}. Let me paraphrase an example from astrophysicist Hugh Ross to help understand this number. Imagine you are in the Superdome at the center of the field. Place one dime on the fifty-yard line. Now imagine the entire field covered in dimes up to a depth of ten feet. Now imagine that the depth of dimes goes to the roof of the Superdome. Now imagine the entire state of Louisiana covered in dimes up to the depth of the height of the Superdome. Now imagine the entire North American continent covered in dimes and to a depth, not of the height of the Superdome, but to a height of 239,000 miles, the distance from here to the moon. Are you still with me? One last step, add one billion more continents covered in dimes to a depth of 239,000 miles. That is your total number of dimes. One of all those dimes is red. Now blindfold a friend and ask him to pick out the one red dime. The odds he will do so are one in 10^{37}.

We could continue to tell of the balance involving the strong nuclear force, the balancing of gravity and the electromagnetic force, the precision needed to balance the electromagnetic force, and the ratio of electron mass and neutron mass to proton mass, and the balances needed for the weak nuclear force, the balances for stars to exist, for carbon to form, this goes on and on. All these different factors and elements have to be "just so" or else there is no universe, and you are not reading this book. Contemplating all this is a dizzying experience. But we will stop here. I only want to give you a taste. Overman notes that these kinds of odds also apply to biology, and the odds are not encouraging for those who believe that the universe and life are the result of totally unguided processes.

Sir Fred Hoyle and Chandra Wickramasinghe calculated the odds of the appearance of the first cell. Assuming 2,000 possible enzymes would be needed for the formation of a first simple cell, the odds of its appearance from random processes came to one in $10^{40,000}$. For help with this number Overman notes that the total number of atoms in the observable universe is 10^{80}.

Confronted with such odds, what would be an appropriate response? Could we grant that it would be appropriate for a person examining such evidence to allow for the possibility that some form of intelligence might possibly be involved? We consider this question in order to contrast the reaction of many physicists to evidence of design in the universe with that of so many biologists.

Biologists committed to Darwin's theory have responded to books like Phillip Johnson's *Darwin on Trial* and Michael Behe's *Darwin's Black Box*, with phrases like, "this will be the end of science," "they do not understand how science works," they have "rejected science," or they "have given up on science," or "the future of science education is now threatened," and this could be "the death of science." The death of science?! Behe has noted the challenge complex biological systems pose for unguided natural processes and has postulated that the evidence points to design. Could this kind of thinking lead to the "death of science?" Now it is possible that Dr. Behe is wrong, but is it reasonable to use this kind of language because someone questions Darwinian orthodoxy in the face of apparent design? Again, let's not forget that until Darwin's theory obtained a lock on scientific thinking in the West, scientists like Newton and others before him had managed to do science quite well believing all the while that the universe was designed.

Here are contrasting quotes from physicists, confronted with evidence of design. The contrast is striking:

Astronomy leads us to a unique event, a universe which was created out of nothing, one with the very delicate balance needed to provide exactly the conditions required to permit life, and one which has an underlying (one might say supernatural) plan.

—Arno Penzias, Nobel prize winner in physics

We are, by astronomical standards, a pampered, cosseted, cherished group of creatures, if the universe had not been made with the most exacting precision we could never have come into existence. It is my view that these circumstances indicate the universe was created for man to live in.

—John O'Keefe, NASA astronomer

I find it quite improbable that such order came out of chaos. There has to be some organizing principle. God to me is a mystery but is the explanation for the miracle of existence, why there is something instead of nothing.

—Alan Sandage, winner of the Crawford prize in astronomy

There is for me powerful evidence that there is something going on behind it all. It seems as though somebody has fine-tuned nature's numbers to make the Universe. The impression of design is overwhelming.

—Paul Davies, astrophysicist

Amazing fine-tuning occurs in the laws that make this (complexity) possible. Realization of the complexity of what is accomplished makes it very difficult not to use the word 'miraculous' without taking a stand as to the ontological status of the word.

—George Ellis, British astrophysicist

And a final quote from British Astrophysicist Sir Fred Hoyle:

A common sense interpretation of the facts suggests that a superintellect has monkeyed with physics, as well as with chemistry and biology, and that there are no blind forces worth speaking about in nature. The numbers one calculates from the facts seem to me so overwhelming as to put this conclusion almost beyond question.

It is noteworthy that so many of them respond to the evidence of design so much differently from so many of those in the biological sciences. I particularly like Sir Fred's phrase "a common sense interpretation." Common sense. We nonscience types who cannot keep up with a molecular biologist or astrophysicist are encouraged by that kind of language. A common sense interpretation of the facts suggests design. If physicists can acknowledge the evidence of design and retain their standing as reasonable thinkers in their field, should not the biologist? Can the detection of design in biology threaten science as we know it, but not if the evidence is found in the cosmos at large?

When one is confronted with the bombastic rhetorical flood that comes from so many defenders of Darwin, one feels that one is dealing with a nineteenth-century theory preserved in amber that is impervious to challenge. No, not so much impervious to challenge, but that the idea of a challenge is in and of itself some kind of affront. The theory's problems have only increased over time with our increase of knowledge, but many of Darwin's defenders do not sound like *scientists* speaking in his defense. Adrian Desmond's biography of Thomas Huxley, *Darwin's Bulldog*, is subtitled: *From Devil's Disciple to Evolution's High Priest.* How fitting. When defenders of Darwin use phrases like "the end of science," they sound like priests defending religion rather than scientists defending a theory.

CHAPTER 14

LESSONS FROM DNA

The nature of DNA as a text is far more than a metaphor.

—Robert Pollack, *Signs of Life*

LET US LOOK AGAIN AT the world of the cell. In his lifetime Darwin aimed, for the most part, to steer clear of how life began. He allowed only the following speculation: "It is often said that all the conditions for the first production of a living organism are now present, which could ever have been present. But if, (and oh! what a big if!) we could conceive in some little pond, with all sorts of ammonia and phosphoric salts, light, heat, electricity, etc. present, that a protein compound was chemically formed ready to undergo still more complex changes, at the present day such matter would be instantly devoured or absorbed, which would not have been the case before living creatures were formed."

Here he speculates on the primordial soup that could bring about life from nonlife, and it has become a staple of life scenarios ever since. (In grade school science class, we learned that men at one time believed in spontaneous generation, that is that nonlife could generate life. But we were taught that science had shown that only life could bring forth life, and that spontaneous generation was an error. However, if natural processes alone can account for the first living cell without an outside intelligence, then one must believe in spontaneous generation after all!). But life did begin, and it is thought that it began with a one cell organism. This seems reasonable enough. (It would hardly do to propose that the primordial soup belch up lions and tigers and bears.) So, we start with the simple cell. Darwin can perhaps be forgiven for thinking that a cell could emerge from nonlife; for him, the cell was a tiny blob of matter. However, we know better. We know that a cell is not a blob of matter, but we need a worthy picture for our minds in order to properly contemplate what it would mean for nonlife to bring forth this first cell. The following from molecular biologist Michael Denton is just the thing to bring to our minds when we hear the words "a simple cell."

Viewed down a light microscope at a magnification of some several hundred times, such as would have been possible in Darwin's time, a living cell is a relatively disappointing spectacle appearing only as an ever-changing and apparently disordered pattern of blobs and particles which, under the influence of unseen turbulent forces, are continually tossed haphazardly in all directions. To grasp the reality of life as it has been revealed by molecular biology, we must magnify a cell a thousand million times until it is twenty kilometers in diameter and resembles a giant airship large enough to cover a great city like London or New York. What we would then see would be an object of unparalleled complexity and adaptive design. On the surface of the cell we would see millions of openings, like the port holes of a vast space ship, opening and closing to allow a continual stream of materials to flow in and out. If we were to enter one of these openings we would find ourselves in a world of supreme technology and bewildering complexity. We would see endless highly organized corridors and conduits branching in every direction away from the perimeter of the cell, some leading to the central memory bank in the nucleus and others to assembly plants and processing units. The nucleus itself would be a vast spherical chamber more than a kilometer in diameter, resembling a geodesic dome inside of which we would see, all neatly stacked together in ordered arrays, the miles of coiled chains of DNA molecules. A huge range of products and raw materials would shuttle along all the manifold conduits in a highly ordered fashion to and from all the various assembly plants in the outer regions of the cell.

We would wonder at the level of control implicit in the movement of so many objects down so many seemingly endless conduits, all in perfect unison. We would see all around us, in every direction we looked, all sorts of robot-like machines. We would notice that the simplest functional components of the cell, the protein molecules, were astonishingly complex pieces of molecular machinery, each one consisting of about three thousand atoms arranged in highly organized 3-D spatial conformation. We would wonder even more as we watched the strangely purposeful activities of these weird molecular machines, particularly when we realized that, despite all our accumulated knowledge of physics and chemistry, the task of designing one such molecular machine—that is, one single functional protein molecule—would be completely beyond our capacity at present and will probably not be achieved until at least the beginning of the next century. Yet the life of the cell depends on the integrated activities of thousands, certainly tens, and probably hundreds of thousands of different protein molecules.

We would see that nearly every feature of our own advanced machines had its analogue in the cell: artificial languages and their decoding systems, memory banks for information storage and retrieval, elegant control systems regulating the automated assembly of parts and components, error fail-safe and proof-reading devices utilized for quality control, assembly processes involving the principle of prefabrication and modular construction. In fact, so deep would be the feeling of *Deja-vu*, so persuasive the analogy, that much of the terminology we would use to describe this fascinating molecular reality would be borrowed from the world of late twentieth-century technology.

> What we would be witnessing would be an object resembling an immense automated factory, a factory larger than a city and carrying out almost as many unique functions as all the manufacturing activities of man on earth. However, it would be a factory that would have one capacity not equalled in any of our own most advanced machines, for it would be capable of replicating its entire structure within a matter of a few hours. To witness such an act at a magnification of one thousand million times would be an awe-inspiring spectacle.[69]

---

69 Michael Denton, *Evolution: A Theory in Crisis*, Adler & Adler, Chevy Chase, MD, 1986, p. 328, 329.

Now whenever you hear anyone casually remarking about chemical processes on their own getting life started you will have, as it were, a talisman to protect you. You can bring to mind Michael Denton's cell, big as an airship, and recall some of the details just enumerated. Some writer will state that science has not quite figured out the details, but with confident tone the writer will assert that science is sure that, without the aid of any outside agency, natural processes somehow got the job done. Thanks to this picture you will have been vaccinated against that easy confidence. Even if at this point, if one is still confident in today's version of evolution, I believe the objective observer will at least appreciate why some today have become skeptics, and that the skeptic's doubts are rooted in what modern science is revealing. This skepticism can be a comfortable fit for those who believe in God, and those who do not. The skepticism is not based on some particular religious framework. It results from honestly looking clearly at the facts.

Let's take just a moment to consider the significance of DNA (For those who want to explore more see the Addendum at the end of this book). According to Robert Pollack in *Signs of Life, The Language and Meanings of DNA,* the 1969 edition of the Encyclopedia Britannica contains about two hundred million letters. The human genome, your DNA, contains about three billion letters. The word *awe* is much overused today. Your encyclopedia might take up two shelves of a moderate-size bookcase for some two hundred million letters. In one of your cells, an object that can only be seen by an electron microscope—three billion letters. Now *that* is awesome! (By the way, Denton's airship cell was magnified, but if we stick with an unmagnified molecule of DNA and unravel it in one straight line it would be six feet long!)

Here from Pollack's book is a small sample of what DNA does because it works like our alphabet: "In human beings, as in every other living thing, DNA tells each cell exactly how to produce the thousands of other molecules that maintain the cell's shape and its place in the body. It is DNA that tells a liver cell to stock the blood with fresh proteins and to store sugar, a nerve cell to stretch itself out into long threads whose tips communicate with the others they touch…the DNA of a cell contains the instructions for making precise copies of itself and of the cell it is in. Each of us has always had, in each of our cells, a DNA text that guided our development from fertilized egg to embryo, fetus, and person; it is a precise copy of our sole and complete inheritance, one that is far more ancient than any human artifact."[70]

Did you notice the words Pollack uses like "tells" and "text"? Elsewhere he speaks of the "linguistic properties, syntax, grammar, and semantics of the human genome."[71]

Here again from Pollack: "The cell's language is grammatical, with a simple syntactical structure. Genes, the stretches of DNA that are capable of speaking through the proteins they encode, are divided into words, called domains, which are quite like verbs and nouns. Just as an English sentence must have a noun and a verb, so proteins must have at least two functional parts. Verb domains ("do this") convey the specific action a protein will take, while noun domains ("to that") convey the target of the protein's action. The minimal protein domains for "do this" and "do that" may be given greater specificity—or more subtle meanings—by other protein domains."[72]

We are starting to get the picture. If the number of letters in DNA was awe-inspiring, how can we find language equal to the task of acknowledging what modern science shows us goes on in the cell? Information is conveyed, instructions given, commands obeyed!

In 1860, Darwin's theory faced formidable hurdles from science. He acknowledged he was asking a lot of his audience to believe that natural selection could do all he believed that it could. We took time to look at why so many of his contemporaries could not grant that he had shown it was so. The discoveries of twentieth-century science have only compounded the problems for evolution. The much-needed evidence from fossils never materialized. The advances made possible by the electron microscope and in cosmology have only added hurdle upon hurdle. Richard Dawkins has said that "It is absolutely safe to say that, if you meet somebody who claims not to believe in evolution, that person is ignorant, stupid or insane (or wicked), but I would rather not consider that." Would not an honest and objective observer of the debate disagree with Dawkins's fanaticism? One can evaluate the evidence and

---

70 Robert Pollack, *Signs of Life*, Houghton Mifflin, Boston, New York, 1994, p. 17.

71 Pollack, *Signs of Life,* p. 68.

72 Pollack, *Signs of Life,* p. 72, 73.

still think the scales are in Darwin's favor. But a fair response to today's evidence is to allow skeptics the legitimacy of their skepticism. If one is not blinkered by the out-of-date idea that a literal reading of Genesis sparks the debate, but rather that it is the works of modern science itself that are the cause for doubt, then one knows that the skepticism is based not on religion. It is based on science and is fully warranted.

CHAPTER 15

FINAL THOUGHTS

In our look at the triumph of Darwinian evolution, I suggested that Hans Christian Anderson had something to teach us about its success. The critique from the following title suggests another mythic story, this one turned to film: *The Wizard of Oz.* You recall that after reaching the Emerald City, Dorothy, Toto, and her three companions are finally in the presence of the "great and powerful Oz." Smoke, fire, and the booming voice, all have the little group shaking in fear. They stand little chance of having their requests granted from the scary Oz. But rescue comes from the little dog Toto. Toto pulls back a curtain to reveal an ordinary man using technology to make himself out to be something he is not. With the curtain pulled back and the ordinary man exposed, Dorothy and her friends are no longer frightened. They can approach him and have a conversation among equals. Without all the huffing and puffing, he cannot intimidate.

We have all been taught since childhood that we must acknowledge the truth of Darwinian evolution. We have been taught that it holds, not only the key for understanding biological life but the key for understanding who we are as human beings. A mighty tall order that . Because it is "science" speaking, we have all thought that, no matter how great our doubts, they could not compete with this "great and powerful Oz."

If you take the plunge that the critics of Darwin offer you, you might find you are like Dorothy, rescued by Toto. What you thought was a theory fully demonstrated and backed with hard scientific evidence is more like a belief system. Once the curtain is pulled back you may see that Darwinian evolution is not nearly as scary as you thought. Beyond what is plain for all to see—that species undergo change—the grand claim that natural selection, in whatever formulae it is presented, can account for the origin of all species is a claim unsubstantiated.

With evolution exposed, like the man behind the curtain, you no longer need be intimidated by the booming voice that says, "Science has shown evolution to be true, you must believe or be counted among the ignorant or insane!" Like Dorothy you will find yourself free—free of fear, and free to listen to your heart as well as your mind.

ADDENDUM

PART ONE:

THE CAMBRIAN EXPLOSION, DNA, AND THE PROBLEM OF INFORMATION

The Cambrian explosion, when new life-forms appeared suddenly between 543 million to 490 million years ago, was a problem that Charles Darwin acknowledged. A problem that he admitted, if unresolved, would prove fatal to his theory of transmutation of one species to another. These new animals appeared suddenly without any transitional forms preceding them. The "innumerable" transitional forms that should have preceded these animals have never been found. By the light of his theory one species evolves through a great number of transitional forms to reach a new and different species. Unfortunately for Darwin, the fossil record, as we have already noted, by itself, undermines the theory today as it did in 1859. Questions we might ask today: did the Cambrian explosion turn out to be "fatal" to his theory of "transmutation by natural selection"? Over the years have advances in science mitigated this problem? As our knowledge increased, have advances in science taken the sting out of the dilemma, or have we found that our knowledge has in fact worsened the sting?

Stephen Meyer's book *Darwin's Doubt* details how the discoveries of modern science have greatly increased the likelihood that the problem is fatal. The primary reason for this is because of what we have learned about the relationship of information to living organisms. This may seem a bit strange at first, but this realization of the central role of information in living things, and in the cell, begins to make sense once one becomes familiar with what was revealed beginning in 1953.

In 1953 Francis Crick and James Watson shared with the world the structure of the molecule deoxyribonucleic acid or DNA. The four bases of the molecule, (cytosine, C; guanine, G; adenine, A; and thymine; T) were shown to be hooked together in the now famous double-helix formation. Their discovery answered some questions in biology but also prompted more. DNA is known to be the source of genetic instructions for building an organism. It is code and instructions for building proteins in living things. DNA is found in the nucleus of the cell. There the information that DNA carries makes it possible to assemble proteins from amino acids. Proteins are the building blocks of life. Different proteins are used to build the different parts of cells and therefore an organism. DNA can instruct to make proteins for blood cells, or muscles cells or bone cells. As the implications of DNA's role in life began to sink in, biologists came to realize that life itself meant a world where instructions were given, received and acted upon. With this realization the discipline of biology had entered a new age, one in which information was paramount. Now what is so special about information?

Let's consider why this is so revolutionary and why the study of living things took on a new caste. But first we need to consider something of the essential nature of information, and how information relates to the empirical world, and the role of the four bases C, G, A, T of DNA. To do that, we have to clearly see how information is separate from any medium that conveys it. Imagine that you teach at a local high school, and you drive to work on a foggy morning. You have a recent model car, one that has a feature that will turn your car lights off if you forget to do so. You arrive in the fog, and as you park, you notice a colleague's car in the slot next to yours. He has a restored 1963 Karmann Ghia. His car does not have that automatic feature. If he forgets to turn off his lights, his battery will run down, and he forgot. You need to alert your colleague so that when he heads home at the end of the day the battery will have the juice to start his engine. You go to his room, but he is not there. You have to leave him a message. Now, you could go to the whiteboard in his room and using the felt pen write, "Bill, your car lights are on." Or you could take a piece of paper from his desk and write your message on the paper using your pen. Or you could go to your computer and send Bill an email warning him that his lights are on and that his battery is in danger of dying in so many hours. Now, when we consider the three ways of giving the same information to Bill, we can clearly see that the message is of a different domain from the means of delivering the message. The message/information is one thing. The medium, however, is a very different thing. We cannot explain the information conveyed by explaining the whiteboard and the felt pen. Nor can we explain the message by accounting for pen and paper or email messaging. All of those are the media available to tell Bill that he needs to act, but the message came from you, an acting intelligent and purposeful agent. You are the source of the information. If you are writing, then you arrange the letters of the alphabet in a specified order that create the words that make up the message he needs to receive. Going to your PC to send the email, you type those alphabet letters in the same specified order so that he will read, "Bill, your car lights are on." Again, we see in this illustration that the message/information is in a different domain from any of the ways used to inform Bill. Something similar is happening in living organisms, and accounting for the chemical properties of the four bases in DNA does nothing to account for the amazing fact that DNA conveys instructions.

So now back to the animals of the Cambrian explosion. The Cambrian explosion has always been a problem for Darwin's theory. His theory envisions that traits are inherited, that there is then some variation that applies to succeeding generations, that certain traits would be beneficial, that from environmental pressures only the fittest variations would help the favored races to survive, and then given lots of time, that natural selection acting, without a goal or purpose, would preserve the beneficial variations and produce a new species. The Neo-Darwinian synthesis of today, and its modifications, understands that mutations occur in gene reproduction and these mutations are the source of the beneficial variations. This is all a very tall order. Can a blind purposeless process really create a new animal, or new features of an animal? Can this process of natural selection build an eye and the nerves to send the visual signal to the brain by a stochastic search?

As difficult as this might appear for a blind process, our gain in scientific knowledge at the cellular level has made this problem worse. Much worse. Thanks to what science has revealed in the last seventy years about the cell, and the workings of biology at the micro level, the Darwinian process faces a nearly insurmountable hurdle. Neo-Darwinism has no way to account for information, and information is essential. Nature working unaided cannot produce information. Information from DNA means that the bases of the DNA are in a specific arrangement, just as you arranged letters and words in a specified order using letters of the alphabet to warn Bill. How does it come to be that, in the long strand of nucleotide bases, they are arranged in just such a way to convey needed instructions or messages for protein assembly? Looking at the natural workings of chemistry alone, one could look forever and not receive an answer.

To review, we have the unanswered problem for Darwinian evolution with the sudden appearance of complex animal forms without the required precursors. We should have a record of transitional animals bridging the simple ones, to the more complex ones that appear later in the fossil record. The animals of the Ediacaran strata that precede the animals of the Cambrian for the most part do not have a head, mouth, gut or sense organs. They lack bilateral symmetry. The ones found in the Cambrian strata have all these features. Darwin understood this jump from animals so simple to ones with those added features, if left unexplained, would be fatal to his theory.

But that is not the half of it. Now we know that *information* is required to get these new body plans. It is not enough to say that natural selection working with beneficial mutations creates new species. If you want to build anything new in biology, you *must* have instructions to do so. You need instructions to build any new feature. Instructions from DNA to build proteins that can be assembled for the new cells needed for the new feature. You want a new animal? Well then, you need lots and lots of instructions to get the job done. These instructions cannot be explained by reference to chemical composition of the double helix. Recall your need at school to get information to your colleague. Accounting for the makeup of the different media you could use to inform your friend does *not* account for the information that originates with you. We know that DNA has the four bases to work with, but it is the *specific arrangement* of the bases that allows it to convey information. Where does the information/the message come from? How is it that DNA has a code of instructions for building at the molecular level? This part of biology is now one of the great hurdles that Darwin's central idea must overcome. Now we have no experience of information resulting from natural processes, none. But we are not without an answer to our problem. Our experience tells us that information comes from an intelligent agent. The poetry of Shakespeare came from his mind. Huckleberry Finn came from the mind of Mark Twain. Beethoven's Fifth came from the heart and mind of Beethoven. The computer program came from a computer programmer. You created the message for your colleague whose car lights were on. So, we are not clueless about the source of how information is generated. Our own experience is the clue—information comes from an intelligent agent of some kind.

PART TWO:

WHY MILLIONS OF YEARS IS NO SOLUTION (THE MATH PROBLEM)

ONE OF THE FACETS OF Darwin's ideas that have gripped the imagination ever since he published is his reliance on time to get from one species to another. Lots of time. If we have lots and lots of time, then no matter how high the hurdle natural selection will get the job done. Millions of years that allow his blind process to achieve what at first glance may seem impossible. Even though there is no guiding intelligence, it is claimed, with enough time, natural selection will winnow out the useless or harmful innovations and hit upon a winner. Today's Neo-Darwinian synthesis posits that mutations in the genes that code for proteins are the source of the innovations that over time create a new species.

So, is it reasonable to assume that Darwin's blind process has enough time to find, by chance, the new information to build a different animal? It is not uncommon while cruising the Internet to run across an article that says scientists believe that birds evolved from reptiles. The article tells you that ancient reptiles have features in common with Archaeopteryx, an ancient bird-like creature, and that birds evolved along this line. *Hmm*, you say, *that's reasonable enough.* But what if you consider what must transpire at the cellular level for such an evolution to occur. Let's just consider one change. Scales have to become feathers. You now know instructions/information must be provided to make all the needed changes at the level of the cell. So how likely is that to occur at the cellular level, with a blind process that must hit upon the successful needed instructions, to build the new cells out of the new proteins from the new arrangement of amino acids?

Bill Gates has said, "DNA is like a computer program but far, far more advanced than any software ever created." Think about that for a moment. DNA is like computer code, but far more advanced. So you want to change scales to feathers, you need new code from DNA to do so. At Microsoft, if the engineers want to give new function to an application, they do not rely on random changes and hope for the best. Any software engineer will tell you that a random change to their digital code results in degrading of function, not new function. Under the Neo-Darwinian synthesis, a random mutation provides the new information to begin building new proteins and then for new cells and new features, or in our, case changing scales to feathers. How likely is that to work? We know in the case of computer code; make random changes, and you degrade the code now operating and you *lose* function, or the system simply crashes; same with a written language. Change four letters in "Bill, your car lights are on" and you have muddied your message. "Bill, your zar lughts arb bn." Random changes do not look like a promising avenue to build something new in biology.

Bill Gates has said that DNA is like a very advanced computer code. Codes transfer information from one domain to another. An example is the binary two-digit computer code that is used for a twenty-six letter alphabet. In computer code 0s and 1s are used in different combinations to represent the twenty-six letters of our alphabet. DNA is directing the assembly of amino acids to form proteins. DNA uses a four-character code to direct amino acids that have a twenty-character alphabet. To begin changing scales to feathers, the lucky mutation must go through the complicated process called gene expression. In this process the double helix must be unwound so that a portion of the information in the DNA can be copied. This part of the process is called transcription and, by itself, involves many steps performed by molecular machines. The second step, known as translation, is even more complicated where further downstream, waiting for our lucky mutation, is an integrated system involving more than one hundred other proteins that then create the new protein needed to begin changing scales to feathers. (If you are now thinking that gene expression itself defies a Darwinian gradual step-by-step process, you would not be wrong. One of the features of our knowledge of the molecular world is the more knowledge we have, the greater the complexity revealed and, therefore, the greater the challenges to a blind step-by-step assembly of new body parts to form a new animal).

We want to return to thinking about odds. What are the chances that we can get the mutation we need to start turning scales into feathers? Or, stated in another way, how likely is it that in searching for a winner we hit instead on a loser? If we are to know something of the odds for success, one of the questions we must ask is: what is the size of the search space? Another way of asking this is to ask, "what are the chances of *not* coming up with the mutation that is needed." The process must find the winning mutation in what number of possibilities, the search space. We are hoping a blind search will hit upon the correct mutation to direct the amino acids to form a different protein, and the amino acids must fold in a very specific way to create a functioning protein—not just any random arrangement of the amino acids will do! For help imagining the problem set for evolution getting the correct mutation, consider the following illustration. I am using the illustration suggested in Stephen Meyer's *Darwin's Doubt.*

Picture a bike: one that is secured by a lock with two dials and ten digits on each dial. You, overcome with a desire for this bike, are going to become a bike thief. You begin by setting the left dial to zero and the right dial to zero. Then you start changing the right dial through the ten digits of the right dial, so you have 0/1 0/2, 0/3, and on and on. Next you set the left dial to one and begin again, 1/0, 1/1, 1/2, 1/3. You begin to see what a large task you have set for yourself. In fact, you have one hundred possible combinations to work through with just two dials each having ten digits (10x10=100). Now notice what happens if the lock has three dials: your possible combinations are 10x10x10 =1,000. The number 1,000 can also be expressed using three as the exponent, 10^3. There are 1,000 possible ways of combining the numbers on the three dials but only one of those 1,000 possible combinations will unlock the lock. We are beginning to see the problem of finding the one workable combination when growth is exponential. If we have a five-dial lock, then our one combination that will unlock it has one chance in 10x10x10x10x10 or 10^5 or 100,000 possible combinations. Imagine if you will, the time it would take you to search for that one combination that will unlock the lock as you search through 100,000 possible combinations. And for a ten-dial lock 10^{10}=10,000,000,000 possible combinations. You now see how exponential growth results in a vast search space of possibilities for that one winning combination.

Now we want to apply this exponential way of thinking to DNA and protein synthesis. A section of DNA that codes for a specific protein is called a gene. The average gene consists of about 1,000 bases of the DNA strand. Three bases from a strand of DNA in a group are needed to designate one of the twenty amino acids. If our gene has 1,000 bases, then it would code for a little more than three hundred amino acids. Our bike lock search problem is now going to pale given the enormous search space found in the biology of the cell. Because twenty amino acids are used in a specific arrangement, and the one gene can code for more than three hundred of them, our functioning protein represents one possible combination out of a staggeringly large number of possibilities, 20^{300}. 20^{300} is roughly 10^{390}. This is truly an unimaginably large number. For comparison, there are only 10^{65} atoms in the Milky Way galaxy. If you are committed to the Darwinian mechanism for creating a new species, how confident can you be that you will get the mutation you need given the odds of our lucky mutation being 1 in 20^{300}? Consider also that turning scales to feathers is only one of the many body changes that must occur to change a reptile to a bird and the many body changes must all be coordinated. One lucky mutation would not be nearly enough to keep the evolutionary process going in the right direction.

Consider also that the lucky mutation must be preserved by natural selection. Natural selection cannot produce the needed mutation. That mutation, all by itself, or any group of lucky mutations, must then somehow confer some kind of an advantage for the animal, for if it does not, then natural selection will eliminate it. Natural selection can only retain what gives our animal a functional advantage. How likely is it that the one lucky mutation, one chance in 20^{300}, will confer an advantage that natural selection will retain? Only if the mutation is helpful will natural selection keep it. If there is no advantage to having this one mutation or the many needed, in forming of a new protein, it will be eliminated. I think a fair-minded person looking at the odds of just one in 20^{300} for the needed mutation out of the possible non-helpful mutations occurring, and then hoping that natural selection will retain it, would not bet on this horse, er lizard. The odds of finding the needed mutation in the search space that large is best said to be impossible. Beyond those odds are the problems of the many coordinated mutations needed, and for natural selection to preserve all of them. The following quote is from Meyer's *Darwin's Doubt.*

> One of the ways to estimate the amount of new genetic information that appeared with the Cambrian animals is to measure the size of the genomes of modern representatives of the Cambrian groups and compare that to the amount of information in simpler forms of life. Molecular biologists have estimated that a minimally complex single-celled organism would require between 318,000 and 562,000 base pairs of DNA to produce the proteins necessary to maintain life. More complex singe cells might require upwards of a million base pairs of DNA. Yet to assemble the proteins necessary to sustain a complex arthropod such as a trilobite would need orders of magnitude more protein-coding instructions. By way of comparison, the genome size of a modern arthropod, the fruit fly *Drosophila Melanogaster*, is approximately 140 million base pairs. Thus, transitions from a single cell to colonies of cells to complex animals represent significant-and in principle measurable-increases in genetic information.[73]

In the 1990s Robert Sauer of MIT did experiments to determine what might be the rarity of functional protein folds. Amino acids fold in very specific ways to create a working protein. Depending on the length of the protein fold, his work showed the ratio of functional to nonfunctional to be in the range of 1 in 10^{90} to 10^{63}. Sauer's work prompted Douglas

73 Stephen C. Meyer, *Darwin's Doubt*, Harper One, New York, NY, 2013, p. 163

Axe, a scientist working at the Medical Research Council Centre at Cambridge, to determine more accurately how rare they might be. Again, like the search space for the lucky mutation, the search space is enormous for the functional protein fold. Axe's work showed the chance of a proper protein fold to be 1 in 10^{77}.

PART THREE:

ENGINEERS SHARE THEIR DOUBTS

THESE KINDS OF PROBLEMS FROM the world of mathematics have been plaguing evolutionary theory since the Wistar Conference in 1966. At that time Murray Eden of MIT convened a group of mathematicians, engineers, and biologists to examine the problem of a blind search to evolve a new species with no assist from an outside intelligent agent. The resultant publication titled "Mathematical Challenges to Neo-Darwinian Interpretation of Evolution" expressed great doubt that it could be done. Our knowledge today only confirms that Eden's group was ahead of the curve in spotting the difficulty of a blind search succeeding when the search space is enormous.

In 1859 Charles Darwin believed that the cell was a kind of undifferentiated bit of protoplasm, a blob if you will. Today we know better. Today we know that the cell is more like an automated factory, a thing of extraordinary complexity. Darwin's ideas were formulated against a backdrop of enormous ignorance about the cell, the very basis of life. He simply did not know what he did not know. Today we know that all life is made up of these living cells. It should not surprise us that at the Wistar Institute Conference of 1966, convened by Murray Eden, himself an engineer, that it was engineers and mathematicians who called into question the confidence of biologists in the Neo-Darwinian synthesis. Two engineer jokes:

Two engineering students were crossing campus when one asked the other, "Where did you get such a great bike?" The second engineer replied, "Well, I was walking along yesterday minding my own business when a beautiful woman rode up on this bike. She threw the bike to the ground, took off all her clothes, and said, 'Take what you want.' " The second engineer nodded approvingly, "Good choice; the clothes probably wouldn't have fit."

A priest, a doctor, and an engineer were waiting one morning for a particularly slow group of golfers. The engineer fumed, "What is it with those guys? We must have been waiting for fifteen minutes!" The doctor chimed in, "I don't know, but I've never seen such inept golf!" The priest said, "Here comes the greenskeeper. Let's have a word with him." He said, "Hello George, what's wrong with that group ahead of us? They're awfully slow, aren't they?" The greenskeeper replied, "Oh, yes. That's a group of blind firemen. They lost their sight saving our clubhouse from a fire last year, so we always let them play for free anytime." The group fell silent for a moment. The priest said, "That's so sad. I think I will say a special prayer for them tonight." The doctor said, "Good idea, I'm going to contact my ophthalmologist colleague and see if there's anything he can do for them." The engineer said, "Why can't they play at night?"

Everyone who knows an engineer recognizes the personality. Engineers are supremely practical. A theory is one thing, but an engineer knows that the theory, if it is going to be more than a theory, must work. A theory for a new bridge will support the weight of bridge traffic or it won't. A new design for an airplane will either lift the plane into the air or it won't. The engineers and mathematicians at Wistar thoughtfully considered the theory of a random change in genetic code to produce function and naturally saw a problem. They knew that with any formal language, such as code, random changes are not a recipe for improved or new function, random changes degrade function.

If you follow the story of Darwin's theory over the years you see a pattern. Over and over when problems that a nondirected blind process would have creating a new species are brought to light, the defenders of the theory dodge the problem. Wherewith, there then can never be a real problem with the theory. The theory is axiomatic. Michael Denton has said it well, "The hold of the evolutionary paradigm is so powerful that an idea that is more like a principle of medieval astrology than a serious twentieth-century scientific theory has become a reality for evolutionary biologists."[74] It really does not matter what problems a blind process will have creating the panoply of life as we know it; for the defenders, problems simple cannot exist. They will say that, yes, scientists still debate the finer points of just *how* the theory operates and somehow gets it all done, but not to worry, the theory itself is never in danger. The *how* is *always* debatable, but even with the how always in flux, this has no negative implications for the efficacy of the theory. The following comes from the website for the National Center for Science Education. Their website states they are "the only national organization devoted to defending the teaching of evolution in the public schools." They are aware that modern science is posing significant challenges to Darwin's theory but want to assure one and all that the theory of evolution is not in crisis. The following is from their website, December 2023:

> "There is no scientific debate about the fundamentals of evolution. Life evolves; species descend with modifications from other species."
>
> When defining evolution they write "evolution, simply put is descent with inherited modification."
>
> The website tells us that, "Mutations are changes in the information of the genetic material…this means a change in the sequence of DNA."
>
> And tellingly they say this about mutations, "Mutations can be beneficial, neutral, or harmful for the organism, but mutations do not 'try' to supply what the organism 'needs'. In this respect, mutations are random—whether a particular mutation happens or not is unrelated to how useful that mutation would be."

BioLogos, another organization defending Darwin today, was founded to deal with the perceived conflict between science and Christianity. This organization works to show that there need be no conflict when both are understood correctly. Although one of Charles Darwin's chief aims with his theory was to remove the Christian God from the story of life on our planet, the Christians at BioLogos are defenders of the theory. Unlike the National Center for Science Education, the folks at BioLogos

---

74 Michael Denton, *Evolution: A Theory in Crisis*, Adler & Adler, Chevy Chase, MD, 1985, p. 306

are no longer holding to the traditional Neo-Darwinian synthesis that was formed in mid-twentieth century. On their website they have moved away from that strict model and use the term standard evolutionary theory (SET) to account for mechanisms other than changes in genes over time to change one species to another. Natural selection is now defined as "the confluence of random phenotypic variation, genetic inheritance, and differential reproductive success." The site also notes that some scientists challenge SET with extended evolutionary synthesis (ESS) because they believe that phenotypic variation is not entirely random and genetic inheritance is the sole driver of natural selection. Nevertheless, like that folks at The National Center for Science Education the folks at BioLogos want you to know that a new way of understanding Darwin's theory does not mean that the theory is in trouble. In an essay from their website David Burdett is emphatic, "Let me be clear upfront, the core of Darwinian evolution itself is NOT being questioned. Indeed as Dobzhanksy, echoing Teilhard de Chardin, asserted decades ago, 'nothing in biology makes sense except in the light of evolution.'"

Nothing to worry about what these two organizations say; no matter the difficulties, be assured all is well with evolutionary theory. That problem with a lucky gene mutation is not really a problem because the blind process has other means of overcoming any hurdle modern science reveals.

Here it is appropriate to acknowledge that evolutionary theory does explain certain phenomena in the natural world. Changes within species, bacteria adjusting to our man-made medicines, random mutations effecting minor changes, the length of the beak of the finch, all have a home in Darwin's insights. A fine place to begin to think well about just what Darwinian evolution can accomplish, but also to consider the limits of its capability, is professor Behe's engaging book, *The Edge of Evolution: The Search for the Limits of Darwinism*. Professor Behe is a trustworthy guide. He writes clearly and engages a receptive reader with the many aspects of evolutionary theory. One comes to appreciate Darwin's insights about the natural world, but understand as well how discoveries in science, discoveries that Darwin could not be privy to, make the grandest claims of evolution, claims that are unsupported by the evidence, experiments, and logic, simply unrealistic. The great problem is believing that a blind random process has *creative power;* power to give us butterflies, frogs, hippopotamuses, bats, whales, eagles, tigers, you, your Labrador retriever, Dante's *Comedia*, Aquinas's *Summa*, the Gospel of John, St. Paul's Epistle to the Romans, the plays of Shakespeare, the poetry of Keats, the operas of Mozart, the symphonies of Beethoven, the jazz of Louis Armstrong, the Beatles' *White Album*, Orson Wells' *Citizen Kane*, Kentucky bourbon, and baseball.

The history of ancient life-forms developing cannot be observed. Science has yet to demonstrate speciation, one species evolving into another, in the laboratory. Nevertheless, these folks say, you can have complete confidence that the evolutionary biologists have the game well in hand. But in light of the confidence shown on the above websites, let us take notice of some recent experiments that shed a different kind of light on the problem of one species slowly evolving one from another.

PART FOUR:

MORE HURDLES; THE BAR GETS HIGHER

LET'S NOW CONSIDER PROBLEMS FOR Darwinian theory of another order: the development of the embryo. Let's give the defenders of Darwin the necessary mutation(s) needed to begin to evolve a new animal. The process they envision has had luck of extraordinary dimensions. The required mutations are present in the DNA. Natural selection was able to stumble on the new information for building a new animal and preserve it. We start with a fertilized egg that possesses this DNA, and it divides and the two cells divide again and we are on the path to an evolving creature.

Christiane Nusslein-Volhard and Eric Wiechaus together won the Nobel Prize in 1995 in Medicine or Physiology for their experiments with *Drosophila Melanogaster*, the fruit fly. Today their experiments are known as the "Heidelberg screens." Starting in 1979 at the European Molecular Biology Laboratory in Heidelberg, they worked at producing thousands of mutations in the fruit fly. Their work caused variations on a subset of regulatory genes that guide embryo development. The result of their mutating was dead fruit flies. They died in the larvae stage. This raises the question, if mutations killed the developing creature in the embryonic state, how can evolutionary mutations build a new animal. We have assumed that the embryo's DNA has been lucky enough to have the required new information for building new body parts from the new proteins. But the "Heidelberg screens" involving thousands and thousands of fruit flies resulted in dead fruit flies. Should not an experiment along these lines give the defenders of Darwin pause?

Early on in the development of an embryo there is not a great deal of cell differentiation. As the embryo continues to develop, however, differentiation is exactly what must occur. The embryo must have blood cells, muscles cells, nerve cells, and on and on. In addition, the building of the body with the many different cell types must be done in the proper sequence at the proper time. The work of Nusslein-Volhard and Wiechaus revealed that the process of development is guided by transcriptional regulators (TRs) or transcription factors (TFs). They help determine what cells get built when. They act somewhat like switches in the developmental process, as in: start building these cells now, now stop those cells and build these cells, and so on and so on.

Evolutionary geneticists Bernard John and George Miklos and Yale University evolutionary biologist Keith Thomson confirm that for a mutation to successfully alter the body plan, the mutation must occur early in development before too much development has taken place. If the mutation's effect occurs late in development, it can effect only minor changes. But experiments show that the early mutation is lethal. The process of building a new body is tightly integrated. When dealing with a tightly integrated system, a change early on has consequences for all the subsequent events downstream. Their experiments found that regulatory genes when mutated can no longer correctly build the growing animal. The early mutation disrupts the orderly building of the different cells in development and the developing animal cannot survive.

Geneticist John F. McDonald of Georgia Tech notes that the genes that are variable in natural or wild populations effect only minor parts of the form of the body, while the genes that effect macro-evolution do not vary. This is not surprising because experiments have shown that variations that can affect the body for a new creature, macro-evolution, cause death. So, we do see minor variations, say Darwin's finches with their changing beak size, but we know from experiments that simply extrapolating from minor changes to new body plans is fraught with insurmountable difficulties. So, a "Catch 22": to effect a real body plan change you must start early in the process after fertilization, but if you start early, the process is stopped because the embryo dies.

A related challenge to the blind undirected process is that of visible body parts. In regard to natural selection, we must keep in mind natural selection will only retain a change in an animal that gives to it a functional advantage. It selects for a feature that will help the animal survive in a changing environment. Changes that would be harmful, natural selection, by the logic of the theory, will be eliminated.

Consider our reptile evolving into a bird. Many body parts must be changing in a coordinated fashion, not just scales to feathers. The blind process not only must retain the advantages of the one new feature, but also all the others needed body changes at the same time. Scales are transitioning to feathers, limbs to wings, mouth to beaks, etc., and this coordination must remain on target over a period of millions of years.

A further problem is imagining how the transitional forms could function in any environment. The transitional form is no longer a true reptile but not yet a true bird. A reptile is adapted to its environment and bird to its environment, but the creature that no longer has scales but not yet feathers is not suited for either environment. No wonder those "innumerable" forms that Darwin's theory requires are not there. By the logic of the theory they cannot survive. Natural selection has eliminated this evolving creature because it is not fit for its environment.

It is a simple thing to say that the limb of a reptile can evolve into a wing. But if one considers for a moment what an engineering marvel a wing is, the difficulties for the blind process to create a wing from a leg are staggering. Michael Denton's marvelous book *Evolution: A Theory in Crisis* brings this vividly to our minds in this way:

> Each feather consists of a central shaft carrying a series of barbs which are positioned at right angles to the shaft to form the vane. The barbs which make up the vane are held together by rows of barbules. From the anterior barbules, hooks project downward and these interlock with ridges on the posterior barbules. Altogether, in the flight feather of a large bird, about a million barbules cooperate to bind the barbs into an impervious vane... In many birds the positioning of the feather is maintained by an intricate system of tendons which allow the feathers to twist in such a way that when the wing is raised they open like the vanes

> of a blind, greatly reducing resistance, but close completely on the downstroke, thus greatly improving the efficiency of flight… The stiff impervious property of the feathers which makes it so beautiful an adaption for flight, depends basically on such a highly involved and unique system of coadapted components that it seems impossible that any transitional feather-like structure could possess even to a slight degree the crucial properties.[75] In the words of Barbara Stahl in her *Vertebrate History: Problems in Evolution*, as far as feathers are concerned "how they arose initially, presumably from reptile scales defies analysis."[76]

What happens if we apply this kind of analysis to an evolving reptilian lung? You and I have a lung in which air flows in and then out. We breathe in, we inhale, we breathe out, we exhale. The same for reptiles. Birds, however, have a lung in which air moves through the lung in one direction. Air enters at one point and flows on exiting the lung having traveled in one direction. There is no inhale followed by an exhale. Now in the process of changing from one kind of lung to another, the evolving creature must always keep breathing. Any glitch in this evolving process, when the lung is not a reptilian lung or an avian lung, and evolution stops. It stops because our evolving creature is dead. It is dead because it is no longer breathing.

Let's consider one more hurdle for Darwin's trial and error process. Once again we are dealing with the development of the body of an animal. This hurdle comes from the work of the late developmental biologist Eric Davidson of the California Institute of Technology. In 2001 Davidson was awarded the International Prize for Biology by the Japan Society for the Promotion of Science. He was also awarded the Lifetime Achievement Award from the Society for Developmental Biology and the A.O. Kovalevsky Medal from St. Petersburg Society of Naturalists. Working with molecular biologist Roy Britten the two formulated a theory, in 1969, proposing that gene expression was regulated by some kind of control system. Davidson and Britten did not work with the fruit fly, buy with *Strongylocentrotus purpuratus* the purple sea urchin.

75 David Berlinski, *The Deniable Darwin*, Discover Institute Press, Seattle, WA, 2009, p. 446

76 Michael Denton, *Evolution: A Theory in Crisis*, Adler & Adler, Chevy Chase, MD, 1985, p. 209

We understand that an embryo is set, programmed, to grow from a fertilized egg to a mature creature. Let's consider the beginning of the developing embryo as cells divide, two cells, then four, then eight, and its progression to a mature form that has trillions of cells. Again, all along the way in this process, cells must become more and more differentiated. As it develops, the right cells must appear at the right time. We now know that non-protein-coding parts of the genome regulate the timing of the arrival of the different cells. Davidson and Britten's theory has been confirmed. There are signaling molecules that start and stop gene expression so that, just like the building of a house, the building of the growing animal is built in proper sequence.

This network of signaling is called a developmental gene regulatory network (or dGRN). Throughout the process of the embryo's growth, these networks are ensuring that that body plan of the creature is built in the proper order. Experiments have shown that the dGRNs do not and cannot vary, because if the dGRNs do vary, the proper cells do not appear at the proper time. Furthermore, if the dGRNs experience mutation, the developing creature is destroyed. Imagine builders building a house. They follow a proper sequence. The foundation comes first, then the frame, then trusses, etc. It is not possible to successfully build the house out of sequence, so with the developing embryo. This signaling during development is highly integrated with one signal triggering another then another and so forth. Again, since the network is tightly integrated a change, or mutation, to one part of it wrecks the entire network. Davidson's work has profound implications for how a Darwinian mechanism could alter a body plan. The following is a quote is from Davidson's *The Regulatory Genome: Gene Regulatory Networks in Development and Evolution*:

> Neo-Darwinian evolution…assumes that all process works the same way, so that evolution of enzymes or flower colors can be used as current proxies for study of evolution of the body plan. It erroneously assumes that change in protein-coding sequence is the basic cause of change in developmental program; and it erroneously assumes that evolutionary change in body-plan morphology occurs by a continuous process. All of

these assumptions are basically counterfactual. This cannot be surprising, since the Neo-Darwinian synthesis from which these ideas stem was a premolecular biology concoction focused on population genetics and...natural history, neither of which have any direct mechanistic import for genomic regulatory systems that drive embryonic development of the body plan.[77]

77 Stephen C. Meyer, *Darwin's Doubt*, Harper One, New York, NY, 2013, p. 269

PART FIVE:

SUMMING UP: WHAT CAN WE EXPECT FROM A BLIND RANDOM SEARCH

DARWIN PROPOSED THAT ONE SPECIES evolved from another and another and another, beginning after the first simple forms of life began on earth. He proposed his theory to show that there was no need for a guiding intelligence to oversee the new animal forms. He feared that the sudden appearance of the Cambrian animals could be a fatal bullet to his theory. Because of the advancements in scientific knowledge, we now know something even more damaging to the idea of species evolving by natural selection's work. New features for an animal and/or a new animal require information. They require instructions for building new cells from proteins, for new amino acid folds that are required at the molecular level. The inquisitive observer might ask, "Where does the information come from?" Information is simply not something lying about in the natural world ready to be used. Those four bases in the DNA convey instructions for building proteins because they are arranged in a specific way. Imagine if you grabbed a handful of Scrabble tiles from your Scrabble game box and tossed them on a table. The letters at this point do not provide any message. But you can arrange your letters to spell words and arrange your words in a sequence that conveys a message. In addition, you can rearrange the same letters to say something different. DNA in the pre-Cambrian animals was directing the building of very simple animals that are found in the Ediacaran strata. Suddenly the much more complex creatures appear, and we know that at the molecular level new instructions have to be in operation. Like changing our Scrabble tiles to spell different words for a different meaning, the four bases in DNA must now be in some new and more *complex* and *specific* arrangement and providing more information than before. And if all that were not enough, science shows that the building of a new feature in an animal is fraught with difficulties because the growth of the embryo is directed by a tightly integrated regulatory network that cannot suffer alteration. All of this points to an outside intelligence as the source for the needed information and guidance if a new species is to appear.

The brief look at the odds of a blind process providing this information is something we have explored from Stephen Meyer's *Darwin's Doubt*. The odds of a blind search finding the needed new information are what we should expect given the vast search space. The fact that the space is so large means that it is so much more likely that random mutations will fail to hit upon a winning mutation that leads to effective new information. Murray Eden, the electrical engineer, gathered other engineers and mathematicians to Philadelphia in 1966. These were people who understood how a formal language, such as computer code, works to convey information. Just like the engineers in the stories these were men who think in practical terms, can this theory actually deliver! They were familiar with computer code and how that arrangement of 0s and 1s could be arranged specifically to code for operating systems. They understood the implications of the discovery of DNA, and how it operated. At the Wistar Institute Conference they confronted leading evolutionary biologists with their doubts. Knowing how computer codes get new function, and understanding that random changes in code degrade function, they, in a sense, told their biologist friends, "Fellas, you have a problem." As you can imagine, committed Darwinists didn't much care for the criticism, especially from engineers and mathematicians who did not understand Neo-Darwinian theory. In general, the biologists told their critics to pound sand. But today's scientific knowledge only confirms that the engineers were correct to ask their biology friends how a blind search could deliver the goods.

We have looked at problems for evolutionary biology at the molecular level, as well as problems for the theory in regard to the development of body plans. Earlier we noted that in observing the world we live in we see design. Noted Darwin champion Richard Dawkins states in chapter one of his book *The Blind Watchmaker*, "Biology is the study of complicated things that give the appearance of having been designed for a purpose." Writers like Dawkins have convinced multitudes that Darwin, and those who have followed in his steps, and the discoveries and experiments of science over decades, have proven that there is no designer needed to account for the design that we so easily perceive. But we know from the discoveries of science made in the last seventy years that Dawkins's words "ain't necessarily so." Knowledge of biology at the molecular level shows the clear need for an outside intelligence to account for information. We clearly see that a blind process has little to no chance of generating new information. Nor could a blind process properly reorder gene regulatory networks and other aspects of epigenetic information that direct the building of an animal's body. However, we have our own long experience in life that intelligence can provide information/instructions/messages. The illustration with the Scrabble tiles shows us clearly that the tiles themselves, like the four bases of DNA, are not enough. The tiles or the bases must be arranged in a specified order in order to carry information. There have always been those who questioned whether the process Darwin first laid out had the creative capacity to give us the world of nature that we see today. Those skeptics have been a slim minority, but their numbers are now growing. Many of the new skeptics are coming from the ranks of evolutionary biology, and it is not difficult to see why.

Also, let's acknowledge how difficult it can be for interested readers to decide where the truth lies when thinking about the claims of Darwinism. Many bright, well-educated folks are convinced that the essential claims Darwin made have not been significantly challenged in over one hundred and fifty years. Some of those see his view of biology as essential to understanding not only the history of life on our planet but understanding ourselves and our place in the world. For some, the theory is foundational to all thinking and understanding. The theory permeates all domains and disciplines of life. Many people, for whom Darwin's idea is essential for their own life, will not be able to consider that it needs revision. But others may be open to thinking differently. What I hope our look at the world at the level of the cell shows is that science has not stood still since Darwin's day, and that the challenge to an old argument comes from what science has revealed. Technology has made it possible to explore the cellular world that Darwin did not know existed, and that is making all the difference.

Thomas Kuhn has shown us that challenges to a current scientific paradigm can become problematic. He notes that when there are so many anomalies that the paradigm struggles to account for all of them, then it is time for the paradigm itself to be reconsidered. That time has come for Darwin's theory.

SOURCES

Anderson, Hans Christian, *The Emperor's New Clothes*

Behe, Michael, 1996, *Darwin's Black Box*, Touchstone

Berlinski, David, 2009, *The Deniable Darwin & Other Essays*, Discovery Institute Press

Browne, Janet, 2002, *Charles Darwin: The Power of Place*, Alfred A. Knopf

Browne, Janet, 1995, *Charles Darwin Voyaging*, Alfred A. Knopf

Coyne, Jerry A., 2009, *Why Evolution Is True*, Viking Adult

Darwin, Charles, 1958, *The Origin of Species*, Mentor

Dawkins, Richard, 1996, *The Blind Watchmaker*, W. W. Norton Company

Dembski, William A., 2007, *No Free Lunch: Why Specified Complexity Cannot Be Purchased Without Intelligence*, Rowman & Littlefield Publishers, Inc.

Dembski, William, *Still Spinning Just Fine: A Response to Kenneth Miller*, First Things, December, 2005, No. 158

Dembski, William A., 2006, *The Design Inference: Eliminating Chance Through Small Probabilities*, Cambridge University Press

Dembski, William A., 2004, *Uncommon Dissent: Intellectuals Who Find Darwinism Unconvincing*, ISI Books

Dennett, Daniel, 1986, *Darwin's Dangerous Idea*, Touchstone

Denton, Michael, 1986, *Evolution: A Theory in Crisis*, Adler & Adler

Desmond, Adrian & Moore, James, 1991, *Darwin*, W. W. Norton & Company

Eiseley, Loren, 1961, *Darwin's Century*, Anchor Books

Eldredge, Niles, 2000, *The Triumph of Evolution and the Failure of Creationism*, W. H. Freeman & Co.

Gould, Stephen Jay, 1983, *Hens Teeth and Horse Toes*, W.W. Norton Co.

Hull, David, 1933, *Darwin and His Critics*, University of Chicago Press

Johnson, Phillip E., 1991, *Darwin on Trial*, Intervarsity Press

Johnson, Phillip E., 1998, *Objections Sustained*, Intervarsity Press

Keynes, Randall, 2002, *Darwin, His Daughter, and Human Evolution*, Riverhead Books

Larson, Edward, 1997, *Summer for the Gods: The Scopes Trial and America's Continuing Debate Over Science and Religion*, Basic Books

Larson, Edward, 2002, *The Theory of Evolution: A History of Controversy*, The Teaching Company

Lawrence, Jerome & Lee, Robert E., *Inherit the Wind*, Bantam Books

Meyer, Stephen C., 2014, *Darwin's Doubt*, Harper One, An Imprint of Harper Collins Publishers

Meyer, Stephen C., 2009, *Signature of the Cell: DNA and the Evidence for Intelligent Design*, HarperOne

Moreland, James Porter, 1994, *The Creation Hypothesis: Scientific Evidence for an Intelligent Designer*, Intervarsity Press

Overman, Dean, 1997, *A Case Against Accident and Self-Organization*, Rowan & Littlefield Publishers, Inc.

Pollack, Robert, 1994, *Signs of Life*, Houghton Miflin

Science, July 17, 1981, Vol. 209, No. 4505

Wells, Jonathan, 2000, *Icons of Evolution*, Regnery Publishing, Inc.

Wells, Jonathan, 2006, *The Politically Incorrect Guide to Darwinism and Intelligent Design*, Regnery Publishing, Inc.

Woodward, Thomas, 2006, *Darwin Strikes Back*, Baker Books

Made in United States
North Haven, CT
04 November 2026

21765031R00102